12 Best Sellers and the Best Buys The UK's best-selling cars in 15 categories, from £5,000 cheapies to £100,000 plus exotics.

16 Touring Cars How do they transform that Vectra from showroom best seller to race-track winner?

18 Introduction to the Catalogue How to get the best from the 81-page Catalogue section.

19 Catalogu[illegible] the most signifi[illegible] by people who've driven them!

100 Data The technical background on all the cars catalogued.

109 New car prices A detailed price guide to all the new cars on sale in the UK.

112 The World's Best Sellers The cars that really pull the punches on the world stage.

"The Guide contains technical data and opinions of individual contributors on each of the cars featured. Whilst every reasonable effort has been made to ensure that the information is correct, neither Express Newspapers plc nor Pedigree Books Limited (or their respective agents) will accept responsibility for any loss or damage suffered by reliance on such information."

Car by Car

Your quick reference guide to the 275 cars in the Guide

Make	Model		Page
Acura	CL		19
Alfa	145		19
	146		19
	155		19
	164		20
	Spyder		20
	GTV		20
ARO	4x4		98
Asia	Rocsta		20
Aston Martin	DB7		21
	V8 Coupe		21
	Virage		21

Make	Model		Page
Audi	A3	*new!*	21
	A4		22
	A6		22
	A8		22
Bentley	Saloon		23
	Continental		23
	Azure		23
BMW	3 Series		23
	5 Series		24
	7 Series		24
	8 Series		24
	Z3		25

Make	Model		Page
Bristol	Blenheim		25
Buick	Century	*new!*	25
	Park Avenue		26
Cadillac	Catera	*new!*	26
	De Ville		26
	Seville		27
Caterham	Super Seven		27
	C21		27
Chevrolet	Malibu		28
	Camaro		28
	Corvette		28
	Lumina		29

Make	Model		Page
Chrysler	Neon		29
	Sebring		29
	Voyager		30
	Jeep Wrangler		30
	Jeep Cherokee		31
	Grand Cherokee		31
Citroen	AX		31
	Saxo		71
	ZX		31
	Xantia		32
	XM		32
	Synergie		32
Daewoo	Nexia		33
	Espero		33
Daihatsu	HiJet		33
	Charade		34
	Fourtrak		34
	Sportrak		34
Dodge	Intrepid		35
	Viper		35
Donkervoort	D8		35
Ferrari	F355		36
	456GT		36
	550M	*new!*	36
	F50		37

Make	Model		Page
Fiat	Cinquecento		37
	Panda		38
	Palio	*new!*	37
	Punto		38
	Bravo		38
	Brava		38
	Marea	*new!*	39
	Coupe		39
	barchetta		39
	Ulysse		32
Ford	Ka	*new!*	40
	Fiesta		40
	Escort		40
	Mondeo		41
	Probe		41
	Scorpio		41
	Maverick		70
	Galaxy		42

Make	Model		Page
Ford (US)	Escort	*new!*	42
	Contour		42
	Mustang		43
	Taurus		43
	Expedition	*new!*	43
Ford (Australia)	Falcon		44
FSO	Caro		98
Hindustan	Ambassador		98
	Contessa		98
Holden	Commodore		44
Honda	City	*new!*	44
	Civic		45
	Accord		45
	Shuttle		45
	Prelude	*new!*	46
	Legend	*new!*	46
	CRV	*new!*	46
	CRX		47
	NSX		47
Hummer			47
Hyundai	Accent		48
	Lantra		48
	Coupe	*new!*	48
	Sonata		49
	Grandeur		49
	Dynasty		49

Make	Model		Page
Isuzu	Trooper		94
Jaguar/Daimler	XJ6		49
	XK8	*new!*	50
Kia	Pride		50
	Mentor		50
	Sportage		51
	Avella		51
Lada	Riva		51
	Samara		52
	Niva		52
Lamborghini	Diablo		52
Lancia	Y10	*new!*	53
	Delta		53
	Dedra		53
	Kappa		54
	Z		32

Make	Model	Page
Land Rover	Defender	54
	Discovery	54
	Range Rover	55
Lexus	GS300	55
	LS400	55
Lincoln	Continental	56
	Mark VIII	56
Lister	Storm	56
Lotus	Elise	57
	Esprit	57
Mahindra	CJ	98
Marcos		57
Maserati	Ghibli	58
	Quattroporte	58

Make	Model	Page
Mazda	121	40
	323	58
	626	59
	Xedos 6	59
	Xedos 9	59
	MX-3	60
	MX-5	60
	MX-6	60
McLaren	F1	61
Mercedes-Benz	C-Class	61
	E-Class	61
	S-Class	62
	SL	62
	SLK *new!*	62
	V-Class *new!*	63
Mercury	Mountaineer	63
MG	MGF	63
Mitsubishi	Colt	64
	new Galant	64
	Carisma	64
	3000 GT	65
	Spacer Wagon	65
	Spacer Runner	65
	Pajero Junior *new!*	65
	Shogun	66
	Space Gear	66
	FTO	66

Make	Model	Page
Morgan		67
Moskvich	2141	98
Nissan	Micra	67
	Almera	67
	Primera *new!*	68
	QX	68

Make	Model	Page
Nissan	Infiniti I30	68
	Infiniti Q45	68
	200SX	69
	Serena	69
	Patrol	69
	Terrano II	70
	Skyline	70
	Rasheen	70
Oldsmobile	Cutlass *new!*	71
	Bravada	71
Peugeot	106	71
	306	72
see Citroen	406	72
	605	72
	806	32
Plymouth	Prowler *new!*	73
	Breeze	73
Pontiac	Grand Prix *new!*	73
	Sunfire	74
	Bonneville	74
Porsche	Boxster *new!*	74
	911	75
Premier	118	99
Proton	Compact	75
	Persona	75
Renault	Twingo	76
	Clio	76
	Megane	76
	Laguna	77
	Safrane	77
	Espace	77
	Sport Spider	78

Make	Model	Page
Rolls-Royce	Flying Spur	78
Rover	Mini	78
	100	79
	Coupe	79
	Cabriolet	79
	Tourer	79
	200	79
	400	80
	600	80
	800	80
Saab	900	81
	9000	81
Saturn	SL	81
Seat	Marbella	82
	Ibiza	82
	Cordoba	82
	Toledo	82
	Alhambra	42
Skoda	Felicia	83
Spectre		83

Make	Model	Page
Ssangyong	Musso	83
Subaru	Justy *new!*	85
	Impreza	84
	Legacy	84
	SVX	84
Suzuki	Swift	85
	X-90	85
	Vitara	85
	Baleno	86
Tata	Ghurka	99
Tatra	700	99
Tofas		99
Toyota	Starlet	86
	Corolla	86
	Carina E	87
	Camry *new!*	87
	Avalon	87
	Paseo	88
	MR2	88
	Celica	88
	Picnic *new!*	89
	Previa	89
	RAV4	89
	Landcruiser Colorado	90
	Landcruiser	90

Make	Model	Page
TVR	Cerbera	90
	Chimera	91
	Griffith	91
Vauxhall	Corsa	91
	Astra	92
	Vectra	92
	Omega	92
	Tigra	93
	Calibra	93
	Sintra *new!*	93
	Frontera	94
	Monterey	94
Venturi	400	94
Volkswagen	Polo	95
VW	Gol *new!*	99
	Golf	95
	Vento	95
	Passat *new!*	95
	Sharan	42
	Beetle	96

Make	Model	Page
Volvo	400	96
	S40/V40	96
	850	97
	940/960	97
Westfield		97
ZAZ	Tavria	99

New for '97

1997 looks like being a better year than most for new models, both in Europe and throughout the rest of the world. The choice runs from the novel - the General Motors Electric car (a genuine production reality) - to the reincarnated VW Beetle, the bizarre Plymouth Prowler and the exciting Porsche Boxster. There's plenty for Mr Practical too - people carriers from Renault, Vauxhall, Opel and Mercedes-Benz. Whatever your needs, there's something new in 1997 to choose from.

Ferrari 550M
Ferrari goes front-engined for the 512M/Testarossa replacement. It's faster, of course.

Honda City
Honda's new 'People's Car', built in Thailand for Asian markets.

Fiat Palio
Fiat's car for developing markets, the Palio will be built world-wide in six body styles.

Lada 2110
Replacement for the Samara is bigger but, crucially, needs to be a lot better too.

Audi A3

The first small Audi for over a decade takes on the BMW Series Compact.

Buick Century

All-new styling for Buick's conservative Century - the previous model ran for 15 years.

Mercedes V-Class

New luxury off-roader from Mercedes-Benz, built in Portugal to keep prices in check.

Mitsubishi Pajero Junior

For those who fancy a Shogun but don't have the room or the budget for a full-sized one.

Jaguar XK-8 convertible

The coupe may look great, but the true star of the XK-8 range is this beautiful new convertible.

Renault Megane Scenic

A new concept? Renault certainly thinks so. Five seats, loads of room, highly versatile, that's the Scenic.

Nissan Primera

It may be great underneath, but is the styling of Nissan's new family car just a little too unadventurous?

Saturn

A new body and interior for the Saturn, General Motors' answer to the cheap imports flooding into America.

Skoda Octavia

Volkswagen (the owner) is gradually moving Skoda upmarket. The Octavia is a full-sized family car.

Toyota Landcruiser Colorado

Replacement for the 4Runner, with safe, Mitsubishi-Shogun inspired styling.

Toyota Paseo
Convertible version of Toyota's cheap and cheerful coupe.

Mercedes SLK
Open-top roadster with clever folding steel roof, all for a budget price - at least by M-B standards.

Volkswagen Passat
VW finally shakes off the steady-but-dull image with the stylish new Passat.

Cadillac Catera
Small Cadillac aimed at the younger buyer, based on the Vauxhall/Opel Omega.

Volkswagen Beetle
After many previews of concept cars, the all-new front-wheel-drive Beetle finally arrives.

GM EV-1

The world's first volume-production electric car goes on sale in California, Nevada and Utah.

Ford Expedition

Enormous nine-seater off-roader with 4.6-or 5.4-litre V8, based on Ford pick-up truck.

Ford Mondeo

Three and a half years on the Mondeo gets a nose and tail job, better security and improved refinement.

Honda CRV

A cross between an off-roader and an MPV, the fun-to-drive CRV arrives in Europe in 1997

Toyota Picnic

'A new motoring concept, the 6-seat Picnic is both versatile and fun' says Toyota.

Oldsmobile Cutlass

General Motors' answer to the ever-popular mid-size Japanese saloons.

Ssangyong Korrando

Korean three-door off-roader based upon Musso chassis; designed by Englishman Ken Greenley.

Pontiac Grand Prix

From concept car to production reality, the Grand Prix comes as a saloon or coupe.

Spectre

Race-car looks, road car practicality from tiny English manufacturer Spectre.

Plymouth Prowler

Chrysler's funky hot rod for the Nineties combines retro looks with modern technology.

Hyundai Coupe

The Korean manufacturer gets serious by proving it can produce exciting cars.

Porsche Boxster

The bedrock of the next generation of Porsche sports cars, the Boxster starts off at a budget £30,000.

Best Sellers

Four Superminis, four Small Family cars and two Medium Family cars. The list, if it were extended, would show more of the same - a solid preponderance of 'ordinary' cars, with only the BMW 3 Series entering the lower reaches of th Top Twenty to offer a little light relief.

New car sales in the UK are dominated by business purchases and businesses want cars which are practical, dependable and cheap to run. Private buyers, of course, also look for the same virtues, but with often less money to spend, sales will be slanted towards cheaper cars - which is where the supermini comes in.

For this section of the Guide we have analysed sales 14 different categories and looked at the best sellers in each and picked out the car which we reckon is the current pick of the bunch. If you are looking for a car which is not run of the mill, pick one out of the eleven 11 which fall outside the supe mini, small and medium family car sector. These cars are not necessarily a better or worse than the best sellers, just different. That's no bad thing, for there are many car buyers, be they the manager of a company fleet of 1,00 cars or a private individual, who are reluctant to buy something out of the o nary. But variety is the spice of life, and without it motoring would be a dul experience indeed.

The UK All-Comers Top Ten

1. Ford Fiesta
2. Ford Escort
3. Ford Mondeo
4. Vauxhall Astra
5. Vauxhall Vectra
6. Vauxhall Corsa
7. Rover 400
8. Peugeot 306
9. Renault Clio
10. Fiat Punto

Budget Buys

1. Skoda Felicia
2. Fiat Cinquecento
3. Rover Mini
4. Renault 5
5. Lada Samara

This category has traditionally presented a stark choice: small economy cars from major blue chip companies on the one hand, bigger offerings from more marginal manufacturers on the other. However, now that Skoda's VW parentage is so evident in the quality of the Felicia, you can just about have your cake and eat it: Escort proportions for less than the price of a Fiesta, Volkswagen fittings and a standard of ride and handling that is right up to the minute. It still has the rattly old pushrod engine but there's a modern VW alternative unit on the way. For many though, the tiny proportions of cars like the Fiat Cinquecento are a large part of their appeal, especially if their main role is one of urban runabout. In this arena the little Fiat stands supreme.
Pick of the Bunch: Skoda Felicia

Superminis

1. Ford Fiesta
2. Vauxhall Corsa
3. Renault Clio
4. Fiat Punto
5. Nissan Micra

Superminis have most of the economy and convenience benefits of the smallest mini class, without the squeeze on accommodation. This makes them perfectly feasible for small families, which is why they outsell the budget class by 40 to one. Ford's Fiesta has traditionally dominated the group despite having several superior rivals. Now though, it's there on merit, the latest versions with new 16-valve engines and revised suspension leading the way in refinement and dynamics. But if interior space is your main priority you would do better looking at VW's Polo or Fiat's Punto. The only new contender in the class is Citroen's Saxo -a reworking of the Peugeot 106.
Pick of the Bunch: Ford Fiesta

Small Family Cars

1. **Ford Escort**
2. **Vauxhall Astra**
3. **Peugeot 306**
4. **Rover 200**
5. **VW Golf**

The biggest market segment of all, it accounts for around a third of all cars sold in the UK. As ever, it is dominated by the Ford Escort, a model past its prime though kept up to acceptable standards by its '95 update. Competing in a market where practicality is everything, cars in this class have traditionally suffered from a somewhat sterile uniformity of design. But all this has changed with the refreshingly diverse forms of Fiat's Bravo/Brava, Renault's Megane, Rover's new 200 and Nissan's Almera. It's hard to pick a winner as each one of these models has its own strengths. It is Peugeot's elderly 306 which still provides the biggest smiles from the driving seat, but the Fiat Bravo is probably the best all-rounder.

Pick of the Bunch: Fiat Bravo

Medium Family Cars

1. **Ford Mondeo**
2. **Vauxhall Vectra**
3. **Rover 400**
4. **Renault Laguna**
5. **Peugeot 406**

The step up to these cars from the class below is substantial in terms of refinement, comfort, quality and equipment. In fact, with the introduction of the Peugeot 406, the best of the class is now touching executive car standards. The 405 replacement has knocked the Mondeo off the critical acclaim pedestal, if not the top of the sales chart. It offers fabulous ride comfort and uncanny suppression of noise and vibration as well as class-leading accommodation. Alongside it, the other new major player in the class, Vauxhall's Vectra, seems remarkably passé. Disappointingly, the Cavalier replacement struggles to reach the standards of the far older Ford Mondeo.

Pick of the Bunch: Peugeot 406

Junior Executive Cars

1. **BMW 3 Series**
2. **Audi A4**
2. **Rover 600**
3. **Mercedes-Benz C-Class**
4. **Volvo 850**

No bigger than the cars from the medium category, these models have an image far removed. Rather than simply doing a job, cars at this level have to say something about their drivers and are priced accordingly. BMW projects the most dynamic image and in the case of the 3-Series backs it up with superb driving qualities, making this the aspirational junior executive car. But the BMW is under threat from Audi's A4, perhaps the most beautiful of these cars and with superb build quality too, though not quite matching the image or the dynamics of the BMW. The Rover 600 and Mercedes C-Class consolidate their positions as the next most popular.

Pick of the Bunch: BMW 3-Series

Executive Cars

1. **Vauxhall Omega**
2. **Mercedes-Benz E Class**
3. **BMW 5 Series**
4. **Rover 800**
5. **Ford Scorpio**

Typically these cars start at around £20,000 and tend to be purchased by companies rather than privately. They're usually available in either four or six-cylinder form - the BMW being the exception with six or eight cylinders - plus a six-cylinder turbo-diesel. BMW's new generation 5 Series has rewritten the rule book here, with fabulous engines, ride comfort and refinement. It could well be the class best seller by the end of the year once the entry-level 520i comes on stream. Meanwhile, Vauxhall's Omega and Mercedes' new E-Class have been making hay, but neither Ford's revised Scorpio nor Rover's new KV6 engine in the 800 have made much of an impact.

Pick of the Bunch: BMW 5 Series

Luxury Cars

1. **Jaguar XJ6**
2. **BMW 7 Series**
3. **Mercedes-Benz S-Class**
4. **Lexus LS 400**
5. **Audi A8**

This is luxury and then some. All these director-level cars will provide quick but near-silent progress, indulgent levels of equipment, cosseting comfort and sheer physical presence. If Jaguar's XJ6 gets some of its feline beauty at the expense of a little leg and headroom compared to the opposition, it's a compromise many in this market are fully prepared to accept. It still leads the field in ride quality, value and, perhaps even more importantly, character. As an all-round package the BMW is arguably more accomplished and its appeal has been strengthened by enlarged versions of its V8 motors.
Pick of the Bunch: Jaguar XJ

Exotics

1. **Porsche 911**
2. **Mercedes-Benz SL**
3. **Jaguar XJS**
4. **Bentley**
5. **Ferrari**

With £50,000 plus to spend on a car, and one which in many cases is too cramped to be considered as an only means of transport, buyers in this market are enjoying pure indulgence and demonstrating serious wealth. There are enough of them in the UK to make this one of the strongest of all markets for the Mercedes SL and Porsche 911, the SL with the emphasis on laid back cruising, the 911 an adrenaline pump. These two will probably have a fight on their hands, though, once Jaguar's XJS replacement, the XK8, hits the air-conditioned showrooms late this year. In the meantime BMW's 8 Series is in decline, with the V12-engined 850 gone, leaving just the 840Ci.
Pick of the Bunch: Porsche 911

Sports Cars

1. **MGF**
2. **Mazda MX-5**
3. **Porsche 911**
4. **Mercedes-Benz SL**
5. **Toyota MR2**

Thanks to the bravery of Mazda in reintroducing the concept of a modern, affordable sports car six years ago, this market has been reborn. The success of the pretty Mazda MX-5 has inspired a whole host of imitators, many of which have yet to hit the UK market. First of the post-MX-5 generation of sportsters is the MGF. This has proved a stunning success, so much so that Rover has been unable to keep up with initial demand. The MX-5 continues to sell strongly thanks to undated looks, driving enjoyment and solid quality. The pretty Fiat Barchetta is only available in left-hand-drive but 1997 could see everything turned on its head with the introduction of BMW's Z3 and Mercedes' SLK.
Pick of the Bunch: MG*F*

Coupes

1. **BMW 3 Series**
2. **Vauxhall Tigra**
3. **Vauxhall Calibra**
4. **Toyota Celica**
5. **Ford Probe**

Despite the presence of two new Latin lovelies in the form of the Fiat Coupe and Alfa-Romeo GTV, BMW's 3-Series Coupe continues to hold sway in the sales charts. Vauxhall's little Tigra has effectively introduced a smaller sub-section into this class, replicating the earlier success of its big brother the Calibra. In the Tigra's wake have come small coupes from both Honda and Toyota. Of the mainstream coupes though, the Calibra continues to outsell its various rivals from Ford, Italy and Japan. Latest in on the act is Hyundai whose slinky new Coupe, with its top-drawer dynamics, should help create a stronger image for the Korean manufacturer.

Pick of the Bunch: Alfa-Romeo GTV

MPVs

1. **Ford Galaxy**
2. **Volkswagen Sharan**
3. **Toyota Previa**
4. **Renault Espace**
3. **Nissan Serena**

Renault more-or-less invented this market many years ago with the Espace. 1996 saw it finally toppled from its tree by the imitators it spawned - at the same time the total MPV market pretty much doubled. Ford's Galaxy (and its VW and Seat clones, the Sharan and Alhambra) combines the versatility of the Espace with a more car-like driving experience. Peugeot's 806 is the best-selling of the identical MPVs from Peugeot, Citroen, Fiat:and Lancia: their design offers the most comfortable ride of the MPV crop, though styling lacks the Galaxy's imagination or the Espace's elegance. But if you need to seat three rows of people and their luggage, the only MPV currently up to the job is the Toyota Previa.

Pick of the Bunch: Ford Galaxy

Off-Roaders

1. **Land Rover Discovery**
2. **Vauxhall Frontera**
3. **Suzuki Vitara**
4. **Range Rover**
5. **Toyota RAV4**

The much speculated drop in off-roader sales hasn't really happened, sales in this category remaining around the same as '95 (albeit within an expanding total car market). The theory went that people would switch from the heavy, thirsty 4x4s whose off-road capability was rarely used to MPVs. But clearly there are plenty who feel that a Range Rover or Shogun carries more prestige than an Espace or Galaxy. Land Rover's Discovery still easily outsells all-comers. In the smaller section Toyota's RAV4 has been given a new rival in the form of Suzuki's X-90, a model which acknowledges how little it's likely to be used off-road by also being available as a two-wheel-drive.

Pick of the Bunch: Land Rover Discovery

Automatics

1. **Mercedes-Benz E-Class**
2. **Mercedes-Benz C-Class**
3. **Ford Mondeo**
4. **Vauxhall Corsa**
5. **Vauxhall Omega**

No longer are you presented with the black and white choice of manual or auto transmission. A combination of conventional automatic gear selection and semi-automatic, selectable at the flick of the lever, is very much in vogue. Pioneered by Porsche as Tiptronic (where you can even have the option of changing by the lever or by F1-like 'paddles' on the steering wheel), similar systems are also used by BMW and Saab. Even conventional automatics are nowadays usually controlled by electronics to give sport or economy modes, with accompanying control of the engine to give seamlessly smooth changes. In lower-powered models, like the Ford Escort and Fiesta, CVT variable ratio transmission is becoming popular.

Diesels

1. **Peugeot 306**
2. **Ford Mondeo**
3. **Vauxhall Astra**
4. **Peugeot 106**
5. **Ford Escort**

Although not quite as popular as in continental Europe, diesels still account for a very significant share of the UK market, with some diesels - notably the Citroen Xantia - comfortably outselling their petrol equivalents. When turbocharged, the best examples of the breed combine near-petrol levels of performance with much superior fuel economy. Leading the field here is VW/Audi with its range of direct injection TDi units, with Rover not far behind (its new 220SDi gives remarkable near-hot hatch performance). But if refinement comes into the equation, Peugeot/Citroen is still ahead of the game.

Pick of the Bunch: Peugeot 406 TD 2.1

From **Road Car** to **Racer**

The car winning races in the Auto Trader British Touring Car Championship may look just like yours, but the transformation of each from road car to racer costs in the region of a quarter of a million pounds. We asked Vauxhall just what it takes.

From bottom left, clockwise:
The Vectra SuperTouring 2-litres, 136 brake horsepower, 1,350 kilograms, £19,375
The body A huge steel cage provides stiffness for the chassis and safety for the driver.

The brakes Massive steel discs drilled for cooling, with up to six callipers instead of two on the road car.
The engine Limited to 2.0-litres with a maximum speed of 8,500 rpm. Power doubled to 300 bhp.
The seat Enormously strong carbon fibre and Kevlar seat pushed towards the centre of the car for optimum balance.

The gears Nothing like the road car. Six gears, sequential change, gearlever mounted on steering column.
The race car 2-litres, 300 brake horsepower, 975 kilograms, £250,000

1997 World Car Guide

The catalogue section which starts opposite contains the low-down on 275 cars manufactured throughout the world. The specification of each has been analysed, technical data gathered and the all-important 1997 developments outlined. And in the majority of cases these cars have been driven by one of the Guide's highly experienced team of motoring journalists, in order to bring you a balanced and informative opinion on what they are like on the road. We think you'll find it as informative as ever.

For each car we have selected a particular model from the range which exhibits the best all-round characteristics. While it would have been all too easy to pick the most expensive model from each range, cost has been included in the equation too, so that our Best All-Rounder represents a fine blend between its qualities as a car and its value for money.

You'll also find reference to the country of manufacture. Today that can mean a car with a 'Japanese' name is just as likely to be built in the UK as the Far East, and the 'British' Ford that is offered in your local showroom could have come from England, Belgium, Germany, Spain, Portugal or the USA. Does it all matter? Probably not, because the manufacturers make doubly sure the new factories they build produce cars to the same or even better standards than the original.

For those of you who demand more in the way of technical data, a comprehensive database is included after the main catalogue section. It contains information on almost every model listed in the main catalogue section, although occasionally data on the newest cars was not available before we went to press.

Finally a comprehensive listing of new car prices for the UK is presented. This section went to press at the last possible minute to ensure it is as up-to-date as possible. However, car prices change on a regular basis, so for the latest up-to-the-minute details look in the weekly Auto Express, which contains the most accurate new car prices in the UK.

FINDING THE CAR

The Car-by-Car listing on pages four and five will help you to find that elusive model. With the motor industry becoming ever more involved with joint-venture manufacturing, it has become increasing difficult to be certain of the origins of a new car. Just take the current crop of multi-purpose vehicles, MPVs or people carriers, call them what you will. Ford is happy for the uninitiated buyer to take its Galaxy as a full-blooded Ford. Volkswagen is the same with its Sharan. The reality is that neither is built in the UK or Germany, as you might expect, but in the same jointly-owned factory in Portugal.

For the Galaxy and Sharan are the same vehicle, jointly developed by Ford and VW in order to half the massive development costs. The only significant differences between the vehicles are the engines, trim and the all-important badges. Now there's even a third version, the Seat Alhambra - its distinguishing feature is the fitment of air conditioning across the range.

In the Guide these cars are treated as one for obvious reasons, so if you cannot find the model you're after look at Car-by-Car for the relevant page number, and below for an explanation of who is doing what.

SMALL CARS

Volkswagen Polo, Seat Ibiza and Seat Cordoba
VW owns Seat, and all these small cars are similar under the skin. The Polo saloon uses the Cordoba body with very few changes

Fiat Panda and Seat Marbella
Fiat owned Seat in the past, with the Marbella a Spanish version of the Panda. Development continued in different directions after the split.

Ford Fiesta and Mazda 121
Ford has a major shareholding in Mazda. The new 121 is a Fiesta in all but name, built on the same production line in Dagenham.

FAMILY CARS

Rover 400 and Honda Civic
Originally a Honda design, but Rover has restyled it and uses its own engines. Both cars are manufactured in the UK.

Volvo V40 and Mitsubishi Carisma
Built in same factory in Holland, with the same underpinnings, but distinctly different bodies and engines. About as different as you can get with a jointly developed car.

MPVS OR PEOPLE CARRIERS

Citroen Evasion, Citroen Synergy, Peugeot 806, Fiat Ulysse, Lancia Z
A joint venture between the French PSA company (Citroen and Peugeot) and Italian Fiat (Fiat and Lancia). Same engines in all, differences in trim.

Ford Galaxy, Volkswagen Sharan and Seat Alhambra
Built in a new Portuguese factory. Turbo-diesel and V6 engines from VW, but each company uses its own 2.0-litre.

JUNIOR EXECUTIVE CARS.

Honda Accord and Rover 600
Another Honda design adapted and developed by Rover for its own needs. Built in different UK factories, the similarities are hidden by subtle changes to exterior.

OFF-ROADERS

Nissan Terrano II and Ford Maverick
Ford bought into Nissan's off-roader at the development stage. Mechanically identical, with minor styling and trim changes.

Isuzu Trooper and Vauxhall Monterey
Sold for several years in the UK by Isuzu and in Europe by Opel, before Vauxhall muscled in during 1994.

US CARS

General Motors
GM builds cars under the brand names Buick, Cadillac, Chevrolet, Geo, Oldsmobile, Pontiac and Saturn. Many are closely related.

Ford
Ford also manufacturers Lincoln and Mercury cars.

Chrysler
Dodge, Plymouth, Eagle and Jeep are all built by Chrysler.

JAPANESE CARS

Toyota
Lexus is the luxury arm of Toyota.

Honda
Honda also sells cars under the Acura brand name.

Nissan
Infiniti is the up-market Nissan range, popular in the United States.

Acura CL

Acura is Honda's US luxury car division, and this sleek, sexy two-door is the first Acura to emerge from Honda's U. plant in Ohio. Loosely-based on an Accord coupe platform, the CL looks its most striking from the rear, with its swooping boot lid and Darth Vader-like tail lights. Power comes from either the Accord's 2.2-litre four-cylinder, or an all-new 3.0-litre V6.

The 3.0-litre CL is the star here. The new 190 bhp V6, which features Honda's wizard VTEC variable valve-timing system, delivers is silky-smooth power with minimal commotion. But this is no sports coupe; auto only for the V6, plus lashings of timber and leather and standard air conditioning and CD, aims the car squarely at the luxury end of the market.

Best All-Rounder: 3.0 CL

BODY STYLES: Coupe
ENGINE CAPACITY: 2.2, 3.0 V6
PRICE FROM: $22,500
MANUFACTURED IN: US

Alfa 145/146

Alfa offers two quite different body styles for its two hatchbacks. The 145 has three doors and a marked 'hot hatchback' look to it. The 146 is the same from the windscreen forwards, but with roomier, more conservative five-door styling. Initially fitted with flat-four engines, these are gradually being replaced in 1996/97 with the new 16-valve straight-four units of 1.6, 1.8 and 2.0 litres. Changes to the suspension and steering are happening too.

While neither of the flat four engines is particularly quick, the first of the new engines - the 2.0-litre - transforms the 145/146 into a delightful sports saloon, so there's hope at the lower end once the new engines arrive. Inside sports seats provide good support, though room in the rear is only so-so in the 146, worse in the 145. The facia is highly stylised, driving position reasonable and the poor gearchange is said to have been improved.

Best All-Rounder: 146 2.0 ti

BODY STYLES: Hatchback
ENGINE CAPACITY: 1.6, 1.7, 2.0
PRICE FROM: £11,500
MANUFACTURED IN: Italy

Alfa 155

The 155 has suffered something of a split personality. Prior to 1995, Alfa's mid-sized saloon never quite made the mark, certainly inferior dynamically to the 164 which preceded it. Then in 1995 it gained lowered suspension with new dampers, and a brand new 150 bhp 2.0-16v engine. For 1996 a new quick steering rack completed the chassis changes, while mid-year the 1.8 engine was replaced by a 16-valve 140 bhp power unit. Also on offer is sports pack which mimics the Touring Car championship winner of 1994 - rear spoiler, side skirts and black 16-inch alloy rims.

The transformation is remarkable. The 155 now turns into corners much more sharply, the gearchange is slick and the new engines are both sweeter and quicker, as well as offering much improved levels of refinement. There is also the option of a 2.5 V6, but the smaller engined 155s are better balanced. Sadly, the driving position is as awkward as ever.

Best All-Rounder: 155 2.0-16 valve

BODY STYLES: Hatchback
ENGINE CAPACITY: 1.6, 1.7, 2.0
PRICE FROM: £15,000
MANUFACTURED IN: Italy

Alfa 164

Eight years is a long time for an executive car to remain broadly unchanged, so the 164 is, not surprisingly, starting to show its age in the face of its competitors, most of which are newer in design. In some ways the Alfa still competes well, not least with its exterior styling which remains a strong point. The engines, too, remain the essence of any Alfa Romeo, with both the 2.0-litre and the 3.0-litre V6 - in either 210 or 230 bhp form - producing all the charisma drivers have come to expect from this sporting Italian breed.

Of course there have been revisior along the way, most recently improve styling and new seats on the 2.0-litre 1996. This Twin-Spark, with a 150 bh four-cylinder engine, retains that Alfa sparkle if not the outright urge of the But the weaknesses are becoming m noticeable, with the notchy gearchang compromised driving position and, in powerful versions, torque steer becon ing less acceptable as time goes on.

Best All-Rounder: 164 2.0 Super

BODY STYLES: Saloon
ENGINE CAPACITY: 2.0, 2.0 turbo, 3.0V6, 2.5TD
PRICE FROM: £18,500
MANUFACTURED IN: Italy

Alfa Spider/GTV

There's no chance of confusing the Spider with the crop of other new two-seaters hitting the market. Tiny headlights, a pronounced wedge shape and a kicked up tail make the Alfa stand out in any crowd. Unlike past Spiders this one is front-wheel-drive, but there's a unique rear suspension design that offers a degree of rear-wheel steering. Engines are either a 2.0-litre Twin Spark 16-valve producing 150 bhp, or, in Europe, a 192 bhp 3.0-litre V6.

Developed alongside the Spider is the hard-topped GTV (also available with a 2.0 V6 Turbo in some countries) which comes with the option of two tiny back seats. Both Spider and GTV drive extraordinarily well, with a memorable bark from the twin-cam engine, quick steering and leech-like grip, although the stiffer structure and suspension make the GTV the more sporting of the two. Thankfully, the driving position is much more to British tastes than earlier Alfas. The boot is minute.

Best All-Rounder: GTV

BODY STYLES: Convertible, Coupe
ENGINE CAPACITY: 2.0, 2.0 turbo, 3.0V6
PRICE FROM: £20,000
MANUFACTURED IN: Italy

Asia Rocsta

Whoever dreamt up the concept of the WW2 Jeep surely could never have imagined that something very similar would still exist some 65 years on. The Korean-built Rocsta may be more brightly coloured than its army counterpart and it has its sights set on fun rather than the enemy but, in essence, there's little difference. Okay, so the Rocsta has a more advanced engine and will even wheeze its way to 80mph – 70mph with 2.2-litre diesel power – but it follows the age-old recipe of a sturdy ladder-type chassis frame with vestigial bodywork and a crude but effective four-wheel-drive system.

It's this last item that gives the Rocsta its appeal. Driving on Tarmac is purgatory. But take it off-road and there's a whole new world of fun to discover. Suddenly, the fact that it's slow, noisy, crude and ill-equipped doesn't seem to matter any more. Fact is, it will go places you thought a mountain goat could never reach.

Best All-Rounder: Rocsta 1.8DX

BODY STYLES: Convertible, Estate
ENGINE CAPACITY: 1.8, 2.2D
PRICE FROM: £9,500
MANUFACTURED IN: Korea

Aston Martin DB7

With the DB7 Aston Martin has moved away from the heavy lines of the Virage and back to the flowing style of the earlier DB Astons, for some still the classic Aston pedigree. And in manufacturing the new car, a whole new approach has been taken. Instead of being hand-built in Newport Pagnell, the steel (rather than aluminium) body is built by a sub-contractor, then the car is assembled at a new factory in Banbury. Much of the mechanical specification is derived from the Jaguar XJS - Ford owns both companies.

To drive the DB7 is as stirring as the looks. The 3.2-litre supercharged engine gives a seamless surge of power, any time it's needed in any gear. The driving position is close to perfection for a sporting car, but the interior is true luxury car. In 1996 there were numerous changes designed to improve on the quality and comfort, as well as the option of automatic transmission and the introduction of the convertible Volante.

Best All-Rounder: DB7 Volante

BODY STYLES:	Coupe, Convertible	PRICE FROM:	£82,500
ENGINE CAPACITY:	3.2 Supercharged	MANUFACTURED IN:	UK

Aston Martin V8

Having been overshadowed by the excitement surrounding the launch of the DB7, 1996 saw the bigger Aston Martin in line for a make-over. Most significantly the Virage gets the bodywork from the high performance Vantage model, a new grille and is re-christened the V8. The 5.3-litre engine is also uprated with cylinder heads and camshafts from the Vantage pushing the power up 10% to 350 bhp. The Vantage continues with its astonishing 550 bhp engine, while other derivatives will follow based around the V8 changes - Volante convertible, Shooting Brake and Lagonda four-door saloon versions.

There's still a place for those who desire a bespoke Aston Martin built in the traditional manner. Hand-beaten aluminium panels, individually assembled engines and a deeply luxurious interior set the car apart, even if the V8 seems big and unwieldy after the DB7. Factory conversions offer ways to make your V8 even more special, including a 6.3 litre engine producing a staggering 500 bhp.

Best All-Rounder: V8 Coupe

BODY STYLES:	Coupe, Convertible, Estate	PRICE FROM:	£140,000
ENGINE CAPACITY:	5.3V8	MANUFACTURED IN:	UK

Audi A3

Audi has moved into the small car market for the first time in recent history with the all-new A3. Golf-sized, but aimed firmly at BMW's 3 Series Compact, it's telling that unlike other Audis the hatchback A3 gets a transverse engine - the chassis will form the basis of the next generation Golf and Seat Toledo too. Initially only the three-door version will be available, but by 1998 five-door and four-wheel-drive options will be offered. In keeping with Audi's prestigious image, there are no small engines - the range starts at 1.6 litre, through to 1.8 and 1.8 turbo; there will also be a couple of turbo diesels.

First reports indicate that Audi has been remarkably successful in developing the A3. The specification levels are high with a very well-finished interior and a high level of refinement on the road. For those in the front, the A3 feels like a much bigger car but not surprisingly, space for those behind is somewhat limited.

Best All-Rounder: A3 1.8

BODY STYLES:	Hatchback	PRICE FROM:	£13,800
ENGINE CAPACITY:	1.6, 1.8, 1.8Turbo	MANUFACTURED IN:	Germany

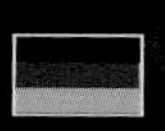

Audi A4

Audi's A4 has quickly progressed into becoming the compact executive car of the late 1990s, taking over the mantle held so long by the BMW 3 Series. For 1996 the range was extended by the addition of the Avant, an estate derivative which is more about style and versatility than serious load carrying. The A4 is well-equipped inside and comes with a wide range of engines, from 1.6 to 2.8 V6 as well as two turbo-diesels. The old Audi 80 continues for the time being in cabriolet form.

Performance from the 5-valve per cylinder 1.8 is impressive, smooth right through the rev range, only lacking verve in fifth gear. The steering and brakes are a little insensitive, but the handling, ride and noise levels are highly satisfactory. Front seats are broad and comfortable but the interior design is clinically efficient and unlikely to win over BMW drivers - the trim materials can be dull, too. Space in the rear is better than a 3 Series, though far from as roomy as mainstream cars like the Mondeo.

Best All-Rounder: A4 1.8 SE

BODY STYLES: Saloon, Estate
ENGINE CAPACITY: 1.6, 1.8, 1.8Turbo, 1.9TD
PRICE FROM: £16,200
MANUFACTURED IN: Germany

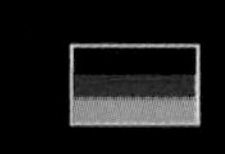

Audi A6

This executive-class range of saloons and estates is strong on safety and has a feel of granite integrity both inside and out. There's a look of understated elegance about the A6, but beneath all the gloss the A6 is now beginning to feel a little jaded dynamically when placed alongside the 5 Series BMWs. This Audi has an excellent 5-cylinder turbo diesel and a punchy V6 to choose from, but the 2.0-litre four isn't really up to scratch. Heading up the range are the sophisticated quattro V6 and the impressively quick S6 with its turbocharged 230bhp engine.

For a large car the Audi's suspension feels harsh on British roads. However, it does tackle the corners with conviction. The saloon is roomy enough and is refined when cruising, but it's the estate that has a more impressive road presence and personality. It's also well thought out and has a load bay roomy enough to cope with most situations.

Best All-Rounder: Audi A6 2.6

BODY STYLES: Saloon, Estate
ENGINE CAPACITY: 1.8, 2.0, 2.2Turbo, 2.6V6, 2.8V6, 1.9TD, 2.5TD
PRICE FROM: £18,900
MANUFACTURED IN: Germany

Audi A8

With the A8 Audi has its most successful stab yet at building a luxury car which stands scrutiny alongside rival marques. The A8 attacks on three fronts: technology, chassis dynamics and style. It looks cleaner and less bulky than its counterparts, and there's no denying a significant technology lead with its all-aluminium body, sophisticated four-wheel drive and an auto transmission that can also be used as a clutchless manual.

The 4.2 V8 engine provides vivid performance, but there's also a 2.8 V-6 and a 3.7 V-8 – the lightest V8 luxury car in the world. These smaller engined models lack four-wheel drive, but they don't really need it. All have a sporting bent with firm suspension, perhaps too sporting for some because the ride inevitably suffers. The Sport models go one step further to create surprisingly crisp responses. Lots of rear space and a comprehensive specification mark this out as a comfortable and fast tourer with continent-shrinking ability.

Best All-Rounder: A8 3.7

BODY STYLES: Saloon
ENGINE CAPACITY: 2.8V6, 3.7V8, 4.2V8
PRICE FROM: £36,000
MANUFACTURED IN: Germany

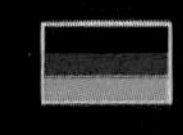

Bentley Brooklands/Turbo R

The Bentley name now stands proud once more, after years of standing for little more than a Rolls Royce with a different grille. The mainstays of the Bentley range are the two saloons, the Brooklands (still available for under £100,000) and the Turbo R, available either in standard form or with a long-wheelbase which allows more rear passenger space. Like all Bentleys, the Brooklands is fitted with a 6.75-litre V8 engine, but for 1997 it receives a light-pressure turbocharger, boosting the power by 25% to 300 bhp - little short of the original Bentley Mulsanne Turbo. The Turbo R has intercooled twin turbochargers which further boost the power to 385 bhp. That's enough to give a top speed close to 150 mph coupled to scalding acceleration.

Both models carry a style that Bentley's one true rival – the S-Class Mercedes – can only aspire to. The Bentley's grandeur and elegance extends to the traditionally handsome interior with its hand-stitched leather and walnut veneer. Bespoke craftsmanship at its best.

Best All-Rounder: Brooklands

BODY STYLES: Saloon
ENGINE CAPACITY: 6.7 V8, 6.7 V8 Turbo
PRICE FROM: £107,000
MANUFACTURED IN: UK

Bentley Continental R/ Azure

There's no more effective way to tell the world you've made it than to travel in extravagant style. The Continental R, or its soft-top stable-mate the Azure, fits the bill perfectly. This top-of-the-range Bentley has a grace lacking from the saloons, with an interior a work of art in leather and wood. The limited edition Continental T takes exclusivity further still, with a shorter wheelbase, more power and aluminium dashboard.

These fast Bentleys are about as sporty as the famous Crewe car-maker gets. That the straight-line urge is there is never in doubt – the twin-turbo 6.7-litre V8 sees to that with its sub-seven second 0-60 mph capability and 150 mph top speed. For swanking around town or for a blast down the autoroute to Nice, there's nothing to match it, but the vast weight tells against it should the driver feel inclined to tackle corners with any commitment. There's more grip than agility.

Best All-Rounder: Continental R

BODY STYLES: Coupe, Convertible
ENGINE CAPACITY: 6.7 Turbo V8
PRICE FROM: £193,000
MANUFACTURED IN: UK

BMW 3 Series

Five years after it was launched, the full range of 3 Series models is only just complete. The original four-door saloon range was quickly expanded to include a coupe, convertible, three-door hatchback - the Compact - and the Touring, a sporting estate. All retain the classic rear-wheel drive design, responsible in no small part for the driving pleasure to be had from a 3 Series. For 1997 the complete range gets a minor facelift, with a fresh grille closer to that in the new 5 Series.

While even the humble 1.6 will entertain, there's a new 1.9-litre 16-valve in the Coupe and Compact which offers a great balance of performance for the money. But it is the six-cylinder models which really impress, from the 2.0-litre to the stunning 192 bhp 2.8-litre, and even the surprising turbo-diesels. The outrageously fast 321 bhp M3 tops off the range. The interior design is aimed at flattering the driver with, apart from a high clutch pedal, high levels of comfort in the front. Rear room is more of a squeeze, however, and equipment levels are not generous
.

Best All-Rounder: BMW 318iS

BODY STYLES: Hatchback, Saloon, Estate, Convertible
ENGINE CAPACITY: 1.6, 1.8, 2.0, 2.5, 2.8, 1.8TD, 2.5TD
PRICE FROM: £13,925
MANUFACTURED IN: Germany

BMW 5 Series

BMW's latest 5 Series aims to move the art of the executive saloon several stages further on. Longer, sleeker and lighter than before, it incorporates a raft of technological developments to improve comfort and performance, yet running costs are also claimed to be significantly lower. Only saloon versions are available now (Touring models are still the 'old' 5 Series), with a choice of six-cylinder engines from 2.0 to 2.8 litres as well as a 2.5 turbodiesel and 4.0-litre V8.

And an extraordinarily impressive car it is. The engines are super-smooth, providing very refined yet at the same time stimulating performance. The 5 Series can be driven almost like a sports car, or will cruise along like a limousine. Seat comfort is good, but although space in the back is much improved it's still some way from roomy. Standard equipment much improved on the new model, including, at last, a stereo.

Best All-Rounder: BMW 523i

BODY STYLES: Saloon
ENGINE CAPACITY: 2.0, 2.5, 2.8, 3.5V8, 4.4V8, 2.5TD
PRICE FROM: £22,500
MANUFACTURED IN: Germany

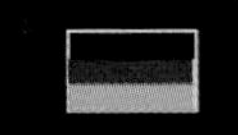

BMW 7 Series

The luxury saloon market is a tough sector for any manufacture, and BMW's 7 Series has to pitch in against the likes of the Jaguar XJ6, Mercedes S-Class and Audi A8. The forte of the BMW is a combination of its unique character and a range of unsurpassed technical developments for the price. Entry level is under £40,000 for the very well-equipped 2.8, but the two new V8s - 3.5 and 4.4 litres cost much more. At the top of the tree is the 5.0-litre V12 which, like the 4.4 V8, can be bought in standard or long wheelbase form.

While the 7 has the luxury leather-and-walnut interior and equipment levels expected in this price range, the V8s lack the ultimate ride comfort or quietness of a Jaguar or Mercedes. That's due to the sporting bias, with delightful performance from the V8s coupled to a highly impressive 'thinking' five-speed automatic transmission. Only the 'servotronic' steering is a disappointment, needing too much twirling around town and lacking feel at speed.

Best All-Rounder: BMW 735i

BODY STYLES: Saloon
ENGINE CAPACITY: 2.8, 3.5V8, 4.4V8, 5.4V12
PRICE FROM: £36,300
MANUFACTURED IN: Germany

BMW 8 Series

Big coupes provide a small but important niche between the luxury saloon and the true supercar. BMW, Jaguar and Mercedes-Benz do these best, with the 8 Series up against tough competition from the new XK-8. Large this coupe certainly is, not far short of two tons in weight, as long as a Scorpio yet providing only enough space behind the front seats for occasional use. Engine choice was initially restricted to a V12, but was later expanded to include a more affordable 4.0-litre V8.

Initially criticised for compromising sportiness in the name of luxury, the CSi, with a 5.6-litre V12 producing an awesome 380 bhp coupled to a six-speed transmission, provided all the performance and handling finesse you could hope for. Market forces, however, have restricted the range to just the 4.0- litre V8 for 1997; with 286 bhp there's still sufficient power to provide a stirring drive - and the luxury is all there too.

Best All-Rounder: 840 Ci

BODY STYLES: Coupe
ENGINE CAPACITY: 4.4V8
PRICE FROM: £56,800
MANUFACTURED IN: Germany

BMW Z3

The last of the current crop of sports cars to hit British streets is arguably the most desirable. For starters the BMW Z3 uses the traditional but increasingly rare mechanical layout of front engine, rear-wheel drive, which promises to give one of the most entertaining driving experiences. Then its styling is unmistakably BMW, with modern touches as well as details which hark back to earlier BMW roadsters, like the vents on the front wings.

It is, of course, the car that James Bond used in Golden Eye. The Z3 is manufactured in South Carolina in the States, where the lower production costs keep the entry price below £20,000. Huge demand worldwide has prevented right-hand-drive Z3s reaching the UK before 1997, by which time the 140 bhp 1.9-litre 16-valve engine will have been joined by a powerful 2.8-litre six; a 300-plus bhp M-Roadster has also been exhibited. Beneath the stylish body the Z3 owes most of its mechanicals to the well-proven 3-Series Compact.

Best All-Rounder: Z3 1.8-16v

BODY STYLES: Convertible
ENGINE CAPACITY: 1.8, 1.9
PRICE FROM: £20,000
MANUFACTURED IN: US

Bristol Blenheim

"A dignified express travel for four six-foot persons and their luggage" is how Bristol describes its Blenheim. Built in the west country factory which originally produced the Bristol aircraft, the manufacturer has been producing bespoke luxury cars since 1947. The key ingredients are a timeless two-door coupe design based around an immense steel chassis and aluminium coachwork.

The 5.9-litre V8 engine and transmission comes from Chrysler in the States, with whom Bristol has a long-standing relationship. Unusual features include wing panels in front of each door to house the spare wheel and battery, which leaves room at the rear of the car for a huge boot. The gearing is so high that the Blenheim will cruise at 70 mph with the engine turning at under 1,800 rpm. While there are similarly-priced cars which have more all-round competence, there's still a place for the endearing Bristol for those who hanker for something distinctive.

Best All-Rounder: Blenheim

BODY STYLES: Coupe
ENGINE CAPACITY: 5.9V8
PRICE FROM: £110,000
MANUFACTURED IN: UK

Buick Century

The long-awaited replacement for the geriatric, old-style Century which, despite its boxy, square edged look, sold over two million copies in its 15-year run. New model is fresh from ground up, with an all-new front-drive platform and well-proportioned, six-passenger body. The smooth-spinning 3.1-litre V6 gets carried over to the new model, though the car's all-coil suspension is new.

Buick is taking no chances here in alienating its existing, somewhat traditional audience. Downy-soft springing, over-light steering and a whisper-quiet engine will appeal to existing buyers, but is unlikely to rake in new ones. Inside, the Century retains its olde-worlde bench-style front seat for those times when families need to carry six in comfort. Excellent value though.

Best All-Rounder: Century Limited

BODY STYLES: Saloon
ENGINE CAPACITY: 3.1V6
PRICE FROM: $17,500
MANUFACTURED IN: US

Buick Park Avenue

Flagship of the Buick line-up, the 'Park' gets a major make-over for '97. Beneath its new skin, it features the extremely rigid platform from Buick's Riviera coupe. Mounted on top is a body that's taller, wider and longer than before - and the previous model was far from small. Standard engine is 205 bhp 3.8-litre V6, while Park Avenue Ultra comes with a potent, supercharged 240 bhp version.

Flagships should be luxurious and the Park Avenue is just that. The giant cabin provides exceptional legroom and headroom, while every comfort known to man is just the press of a button away. Long-travel suspension, hushed engines and one of the smoothest four-speed automatics in the business add to the big Buick's luxury-liner image.

Best All-Rounder: Park Avenue 3800

BODY STYLES: Saloon
ENGINE CAPACITY: 3.8V6, 3.8V6 Supercharged
PRICE FROM: $29,000
MANUFACTURED IN: US

Cadillac Catera

Cadillac desperately needs to draw-in younger, so-called 'baby boomer' buyers. The car that's going to help them is the Vauxhall Omega-based Catera. Built by Opel in Germany, this new Caddy uses the 24-valve 3.0-litre V6 found in both the Omega as well as the Saab 9000. Exterior changes include a new Cadillac-pattern grille and full-width tail-lights, while inside there's a fresh look to the Omega's instruments, less garish woodwork and the US-obligatory cupholders.

Not surprising, the '97 Catera drives just like an Omega. And by normal Cadillac standards, that makes it a road-hugging sports car. The 200 bhp V6 has to work hard to move this hefty four-door, but the four-speed auto is quick to provide kick-downs when needed. With a competitive price tag and an impressive tally of standard fixtures, this new Euro-Cadillac could well bring in that new crop of younger, upwardly-mobiles.

Best All-Rounder: Catera V6

BODY STYLES: Saloon
ENGINE CAPACITY: 3.0V6
PRICE FROM: $29,500
MANUFACTURED IN: Germany

Cadillac De Ville

For '97, Cadillac's luxury liner De Ville gets a more aggressive new look. There's an even wider, egg-crate grille, a bigger power bulge for the bonnet and restyled rear end. The rear track has also been widened to give a more sporty stance. Additional body strengthening also reduces noise inside the car and gives it a more solid feel. Inside, it's all stitched leather and glowing walnut with seating for six and a boot you could hold parties in.

The major improvement for '97 is the addition of Cadillac's so-called Integrated Chassis Control system, which despite the Disneyesque name, really works. Using a combination of anti-lock brakes traction control, variable assisted steering and an array of sensors, the system automatically provides stability on slippery surfaces. It's a worthwhile addition, considering the De Ville now comes with a standard 300 bhp, 4.6-litre Northstar V8 driving its front wheels. Even the base model offers 275 bhp.

Best All-Rounder: Sedan De Ville

BODY STYLES: Saloon
ENGINE CAPACITY: 4.6V8
PRICE FROM: $36,000
MANUFACTURED IN: US

Cadillac Seville

For those searching for that BMW-like combo of athletic performance and entertaining handling, yet prefer to stay loyal to the Stars & Stripes, the Seville is the car of choice. The sporty STS model packs a 300 bhp punch from its 4.6-litre Northstar V8, while the cheaper and smoother-riding SLS copes admirably with 25 bhp less. For '97, both are offered with Cadillac's impressive GPS-based On-Star navigational and emergency rescue system.

Seville earns its colours as an impressive grand tourer. Towering straight-line performance, whisper-quiet high-speed cruising and a relaxed ride make 800 mile-a-day journeys a breeze in this good-looking four-door. But high horsepower and front drive cause the torque-steer blues and the speed-sensing steering is far from linear in its transition from assisted to un-assisted.

Best All-Rounder: Seville SLS

BODY STYLES: Saloon
ENGINE CAPACITY: 4.6V8
PRICE FROM: $43,000
MANUFACTURED IN: US

Caterham Super Seven

1997 sees the 40th anniversary of the incredible Super Seven. Designed by Colin Chapman and built by his Lotus company, production was transferred to Caterham cars in the early 1970s. Since then Caterham has been producing the Seven to an ever enthusiastic market, staying faithful to the exterior design and dimensions but including many technical developments along the way. This year the chassis has been stiffened by 30%, the suspension has been stiffened and there are new seats, instruments and a manageable handbrake for the first time.

Seven buyers can opt for anything from a box of bits to a completely finished car. Engine is the Rover K Series, this year uprated to 1.6-litres in either standard or 138 bhp Supersport form. The 2.0-litre HPC continues for as long as engines are available, while kit car buyers still favour Ford units. Whatever the power, the Seven provides a unique driving experience. With its light weight, low driving position and minimal equipment, this is as close to a racing car for the road as you can get.

Best All-Rounder: Seven 1.6

BODY STYLES: Convertible
ENGINE CAPACITY: 1.6, 2.0
PRICE FROM: £17,900
MANUFACTURED IN: UK

Caterham C21

Two years after it was first shown at the British motor show, Caterham's new sports car is finally making its way into the hands of eager customers. Intended to satisfy owners who have outgrown the Super Seven, the C21 has an all-enveloping body which carries overtones of Lotus sports cars from the Sixties. This time, however, the car will have opening doors - amazingly a first for Caterham. Buyers will have to carry out the final assembly themselves.

The slippery design means a marked improvement on the aerodynamic bluntness of the Super Seven, allowing for higher top speeds although the extra weight will blunt acceleration. The interior is a neat design, readily switchable to left hand drive. Underneath, the C21 is pure Seven, with a wider chassis but similar running gear. The engine is Rover's 1.8 from the MGF, without the variable valve timing but with a few improvements. Bodywork is glass-fibre reinforced plastic, with aluminium a (costly) option.

Best All-Rounder: C21 1.8

BODY STYLES: Convertible
ENGINE CAPACITY: 1.6, 1.8
PRICE FROM: £18,750
MANUFACTURED IN: UK

Chevrolet Malibu

Chevy's all-new Mondeo-sized Malibu aims straight at the heartland of the US car-buying market. Great value-for-money, standard equipment that includes air conditioning, anti-lock brakes and dual airbags and generous interior space should make this new four-door a strong seller when it goes on sale this winter. Styling breaks no new ground but here, buyers care more about price and packaging.

Better-built than most low-cost Chevys, the Malibu has a solid, well-crafted feel to its construction. Noisy and harsh 2.4-litre twin-cam four-cylinder should give gutsy performance, but for those looking for less-frenetic motivation, the optional 3.1-litre, 155 bhp V6 should be the motor of choice. A sweet-changing, four-speed automatic comes standard, as does nicely-weighted power steering. Hard to beat for the money.

Best All-Rounder: Malibu Sedan V6

BODY STYLES:	Saloon	PRICE FROM:	$15,000
ENGINE CAPACITY:	2.4, 3.1V6	MANUFACTURED IN:	US

Chevrolet Camaro

Thirty years young this year, the Camaro still provides the ultimate bang for the buck. For the price of a dull-as-dishwater family saloon, the Camaro Z28 packs a 285 bhp punch from its fire-breathing 5.7-litre V8. Even cheaper - and probably a better all-rounder - is the new 200 bhp 3.8-litre V6 model. Available in coupe or power-top convertible, the Camaro shows that testosterone-charged American muscle cars are still alive and kicking.

Don't expect too much sophistication here. The Z28 is a heavyweight champ that delivers its V8 muscle in an explosion of raw, rumbling, roaring power. Standstill to 60 mph comes up in under 5.5 secs, and with its long-legged six-speed gearbox slotted in top, it'll run to 170 mph-plus. Crude, hunkered-down suspension means the Camaro hops from one pot-hole to the next and the firmness of its ride will loosen dental work. But for the money, it's unbeatable.

Best All-Rounder: Camaro V6

BODY STYLES:	Coupe, Convertible	PRICE FROM:	$15,000
ENGINE CAPACITY:	3.4V6, 5.7V8	MANUFACTURED IN:	US

Chevrolet Corvette

Chevy debuts its all-new 'Vette in January '97, powered by a potent 340 bhp, all-alloy V8. Until then, the current version is still smoking large quantities of tyre rubber. The high-performance Grand Sport packs a 330 bhp kick in the pants from its 5.7-litre LT-4 V8, providing sub-five second sprinting to 60 mph. This optional V8 also comes in the Collector Edition, which has special paint and badging.

There's nothing too subtle about the way this low-slung Chevy delivers its awesome power. Booming noise levels, rumbling road roar and a bone-jarring ride have long been features of America's longest-running sports car. But it is quick. Low roof and high door sills make entry and exit challenging, and once inside the cabin is corset-tight. He-man clutch, heavy brakes and weighty steering justify its 'muscle car' tag.

Best All-Rounder: Corvette Coupe

BODY STYLES:	Coupe, Convertible	PRICE FROM:	$38,000
ENGINE CAPACITY:	5.7V8	MANUFACTURED IN:	US

Chevrolet Lumina

It's tough to argue with great value and Chevy's re-designed Lumina provides big car comfort and equipment levels at small car prices. The styling may be vanilla-bland but the car's long wheelbase and lofty roof line provide a cabin that's light, bright and spacious. The standard hardware includes a hard-grafting, 160 bhp, 3.1-litre V6, with an optional 210 bhp 3.4 for those who want more oomph. Standard kit includes air conditioning, anti-lock stoppers, four-speed auto and dual airbags.

Comfort and refinement is the name of the Lumina's game. Smooth, torquey engines give hushed cruising, while soft, long-travel springing provides a feathery ride. Come to a fast, tight corner, simply slow down, or expect the tyres to squeal and the body to lean. For those who want more of a hint of sportiness, there's a new Lumina LTZ for '97 with the 3.4 motor, 225/60-section rubber and tauter suspension.

Best All-Rounder: Lumina LS

BODY STYLES: Saloon
ENGINE CAPACITY: 3.1V6, 3.4V6
PRICE FROM: £16,500
MANUFACTURED IN: US

Chrysler Neon

The Neon attracts a strong following where it is manufactured, in the States, the friendly, bug-eyed saloon making a lot of friends with its keen price and enjoyable performance. Now it is available in Europe, and from 1996 in right-hand-drive form in the UK. It's pitched at the mainstream family market, but with a 2.0-litre engine and automatic transmission available at no extra cost, its price is more in line with a top-spec Fiesta.

The Europeanised version is fun to drive, with firmer, more sporting suspension and sharp steering. Performance is strong if the engine is extended, but the three-speed auto certainly knocks the edge off. The chief criticism, however, is that while the Neon is comfortable for those in the front, the low front seat cramps up the rear space dramatically - this is truly more an Escort competitor with a big engine.

Best All-Rounder: Neon 2.0 LX

BODY STYLES: Saloon, Coupe
ENGINE CAPACITY: 2.0, 2.0DOHC
PRICE FROM: £11,600
MANUFACTURED IN: US

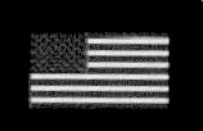

Chrysler Sebring

Despite the badge, this two-door coupe has nothing in common with Chrysler's convertible of the same name. Confused? Although designed by Chrysler, the Sebring two-door is actually built by Mitsubishi at its US plant in Illinois and rides on a modified Galant platform. Power choices include a Chrysler 2.0-litre twin cam 'four', or a Mitsubishi 2.5 V6. Surprisingly, the V6 is only offered in auto form, not five-speed.

The Sebring's aggressive front end, with its slit-like headlamps and deep-set driving lights, screams out performance. But that's the only performance trait that this spacious four-seat coupe shows. Both engines provide lively acceleration, but the four-cylinder sounds harsh and buzzy when revved, and the pace of the V6 is dulled by the auto-only gearbox. But both handle neatly and predictabl, and are easy to live with. They're excellent value too.

Best All-Rounder: Sebring LX.

BODY STYLES: Coupe
ENGINE CAPACITY: 2.0, 2.5V6
PRICE FROM: £16,500
MANUFACTURED IN: US

Chrysler Sebring Convertible

Another star from Chrysler. Here's a full four-seat convertible, with a tight-fitting, fully-lined power roof, great looks with a breezy, fun-to-drive manner for the equivalent of £13,000. A bargain. Designed from the ground-up as a convertible, rather than a coupe with its top chopped, the Sebring is offered as a JX with a Chrysler-made 2.4-litre, 16-valve 'four', or in JXi guise with a Mitsubishi-built 2.5-litre V6.

This is how convertibles should work. Release a catch, press a button and watch as the sunlight pours in. Simple as that. And the heavy lining, glass rear window and double sealing means this is an all-year-round convertible. Despite its length, the Sebring is surprisingly twist and shudder-free; it rides smoothly and remains composed even on the worst blacktop. Four-cylinder motor can become vocal when revved but the V6 is all smoothness and quiet.

Best All-Rounder: Sebring JXi

BODY STYLES: Convertible
ENGINE CAPACITY: 2.4, 2.5V6
PRICE FROM: $19,500
MANUFACTURED IN: Mexico

Chrysler Voyager

With an all-new model for 1996, this Chrysler sets impressive new standards in MPV design. Available in long-wheelbase and short-wheelbase forms, the Voyager comes with a wide array of engines, ranging from a 2.4-litre, four-cylinder, to a 3.8-litre V6. There's even an electric version around the corner. And production is not only centred in the US; Voyagers for Europe are built in Austria and arrive in the UK in right-hand drive form in 1997.

Despite its size, driving the Voyager is like driving a large car. Its power steering is precise and nicely-weighted, cornering is flat with plenty of grip, and performance from the 3.3-litre version is lively and refined. But it is the interior design that impresses most. Sliding doors on both sides make it easy to get in and out, there's plenty of space for seven, even in the short-wheelbase, and the lightweight rear seat glides on rollers for easy removal. The Voyager is destined to be the most practical of all the MPVs on the market.

Best All-Rounder: Voyager 3.3 SE

BODY STYLES: Multi Purpose Vehicle
ENGINE CAPACITY: 2.4, 3.0V6, 3.3V6, 3.8V6, 2.5TD
PRICE FROM: approx. £18,000
MANUFACTURED IN: US, Austria

Chrysler Jeep Wrangler

Taking care not to change the overall size and appearance, the 1997 version of the Wrangler is a thoroughly overhauled vehicle, with three-quarters of the parts replaced or redesigned. New body panels, coil spring suspension, air bags and a move back to round headlamps, like the WW2 Jeeps, are key improvements. The soft top fits better, the rear seat is wider and the windscreen slopes back more - while still retaining that essential fold-flat ability.

Mechanically, the Wrangler remains largely as before, which means four-wheel-drive is, of course, standard, but rear-wheel-drive can be selected for everyday use. Engine choice is between a 2.5 litre, which gives fair performance for this type of vehicle, and a six-cylinder 4.0 litre, giving gut-wrenching acceleration. The Wrangler is easy to drive too, with light steering and great visibility. But even with the myriad of changes, there's no doubting that the Wrangler is more at home in the rough stuff than as a provider of everyday urban transport.

Best All-Rounder: Wrangler 4.0

BODY STYLES: Convertible, Estate
ENGINE CAPACITY: 2.5, 4.0
PRICE FROM: £13,900
MANUFACTURED IN: US

Chrysler Jeep Cherokee/ Grand Cherokee

Offering a blend of American style with remarkably competitive pricing, the Cherokee has proved a strong seller for those looking for an alternative to the Land Rover Discovery. The Grand Cherokee is a larger alternative, with a 4.0-litre, six-cylinder engine, or a 5.2 V8 in left-hand-drive only.

The standard Cherokee may look Discovery-sized, but in reality it is much smaller. The 2.5 engine struggles to cope with the weight, but the turbo-diesel is fine if clattery, while the 4.0-litre is a stormer. The interior design is typically American-tacky, and while comfort levels in the front are satisfactory, rear seat space is surprisingly limited. The Grand Cherokee is an altogether classier, roomier alternative, but slippery leather seats, bumpy ride and modest performance from the 4.0-litre can't be overlooked when the price is close to £30,000.

Best All-Rounder: Cherokee 4.0

BODY STYLES: Estate
ENGINE CAPACITY: 2.5, 4.0, 5.2V8, 2.5TD
PRICE FROM: £17,000
MANUFACTURED IN: US, Austria

Citroen AX

Citroen's AX is fast coming to the end of its days, the new Saxo (see Peugeot 106) filling the mainstream supermini place, leaving just the budget versions of the AX. One of the last-generation superminis, it now finds itself outclassed for space, build quality and safety, but its friendly nature, low running costs and highly competitive pricing enables it to fight its corner. Three or five-door versions are available, with either a 1.0 or 1.5 diesel engine.

Because the AX is a particularly light car, it feels lively and responsive even with a 1.0 engine; the diesel can almost crack 100mph. Around town the AX is a great little car and it copes with the bumps well, but a cramped footwell with offset pedals spoils the comfort for longer journeys. Also, the rear passenger space is cramped, the boot is tin, and it lacks that feeling of integrity now expected in this class.

Best All-Rounder: AX 1.0 Debut

BODY STYLES: Hatchback
ENGINE CAPACITY: 1.0, 1.1, 1.4, 1.5TD
PRICE FROM: £6,300
MANUFACTURED IN: France

Citroen ZX

Citroen's ZX was the first of the Escort-class contenders to prove it was indeed possible to have a practical and roomy five-door hatch that was also great to drive. And while it still possesses fine dynamic qualities, much of the opposition has caught up. An excellent chassis, a generous equipment specification, reasonable space and lively performance are prime assets. But the cabin appears sombre and the 1.8 and 2.0-litre engines are bettered by others. Best of the petrol engined ZXs is the great-value 1.4. An estate is also available, and although it can't quite match an Astra or Golf estate for load capacity, it's nicer to drive.

The most desirable ZX is a diesel, or better still, a turbo-diesel. Citroen scores with its TD not just for impressive economy, but also for fine driveability and better real-world performance than many a petrol engine. There's good refinement too, for a diesel . The ZX is due for replacement towards the end of 1997.

Best Buy: ZX 1.9TD

BODY STYLES: Hatchback, Estate
ENGINE CAPACITY: 1.1, 1.4, 1.8, 2.0, 1.9D, 1.9TD
PRICE FROM: £10,100
MANUFACTURED IN: France

Citroen Xantia

With the Xantia, Citroen has achieved a real pinnacle of automotive achievement. Not only does this hatchback possess infinitely more panache than the average mid-sized high-volume family car, but it also matches or beats most rivals on ability and far more besides. In turbo-diesel form it's just about untouchable, but it's also a front runner with the latest 16-valve petrol engines too.

Part of the reason why it scores over talented rivals is the unique suspension design with its excellent bump absorption properties and stable cornering. But the Xantia is roomy with high levels of seat comfort too. Recent models include the excellent estate (with three full safety belts in the rear), an impressively fast (and impressively thirsty) petrol turbo, the Activa with trick suspension that cuts out body roll in corners and soon, a new V6. In its own way each of the developments is impressive, although Xantias become less convincing as luxury cars when prices approach £20,000.

Best All-Rounder: Xantia 1.9TD LX

BODY STYLES: Hatchback, Estate
ENGINE CAPACITY: 1.6, 1.8, 2.0, 2.0 Turbo, 1.9D, 1.9TD
PRICE FROM: n/a
MANUFACTURED IN: France

Citroen XM

Citroen has never produced a conventional executive car, and the XM is no exception. Not only is the hatchback XM strikingly different to the average model in the executive car park, the magic carpet ride of the air suspension makes it feel very different to drive. An estate also figures in the range, with a vast interior and useful self-levelling suspension. Many XMs are equipped with the economical 2.1 or 2.5 turbo-diesel engine, but there's also a 2.0-16v petrol engine, a punchy 2.0-litre low-pressure turbo and, soon, a new V6.

Introduced in 1989, these days the XM appears dated outside and in, with a facia that's all ledges and edges. However, the vast amount of legroom and load space still impresses, and it corners with fine grip and stability. In turbo-diesel form the XM is capable of more than 500 miles between fill-ups yet it has plenty of pace for fast cruising. A new model is due for 1998.

Best All-Rounder: XM 2.1TD Estate

BODY STYLES: Hatchback, Estate
ENGINE CAPACITY: 2.0, 2.0 Turbo, 3.0V6, 2.1TD, 2.5TD
PRICE FROM: £18,400
MANUFACTURED IN: France

Citroen Synergie/ **Fiat** Ulysse **Lancia** Z/ **Peugeot** 806

This people carrier is the result of a cooperative venture between PSA and the Fiat Group, between them owners of these four car manufacturers. Aimed squarely at the Renault Espace, the Evasion offers a very similar package of six to eight seats, with the rear two rows removable to increase luggage space. Engine choice is between a 2.0-litre and 1.9 turbo-diesel, and in some countries a 2.0-litre turbo and a 2.1 turbo-diesel.

From the passengers' point of view the vehicle is a winner. The interior is extremely well-finished and comfort levels are high with great seats and ride, and low noise levels. Space in the two front rows is good but it's tight for those in the back. From the driver's point of view the 2.0 petrol engine lacks performance (the turbo-diesel is preferable), while the novel dashboard-mounted gearlever can be tiresome in town driving. Like the Espace, luggage space is minimal with all the seats in place.

Best All-Rounder: 2.0 Turbo

BODY STYLES: Multi Purpose Vehicle
ENGINE CAPACITY: 2.0, 2.0 Turbo, 1.9TD
PRICE FROM: £15,900
MANUFACTURED IN: France

Daewoo Nexia

Take a superseded but good-in-its-day small family car, restyle the nose and tail, drop in some modern engines and add a pile of equipment, and you have the Daewoo Nexia, a Korean-built close-cousin of the last-generation Vauxhall Astra. Nothing much wrong with that of course, provided the price is right and the buyers none too discriminating. In fact, it's not so much a car Daewoo is selling here, it's more a complete motoring package. Everything from a servicing contract to a warranty comes as part of the deal. Even safety features are standard; not just an airbag, but anti-lock brakes; there's power steering and air conditioning too.

What you won't get is anything more than mediocrity in the car's dynamics, refinement and packaging. Even so, this range of 1.5-litre hatchbacks and saloons will meet the approval of many. Furthermore, the entire concept could well shake up the motor manufacturers into thinking more about the private customer's needs and less about fleets.

Best All-Rounder: Nexia GLXi 5-door

BODY STYLES:	Hatchback, Saloon	PRICE FROM:	£8,800
ENGINE CAPACITY:	1.5	MANUFACTURED IN:	Korea

Daewoo Espero

Beneath the crisp lines of the Espero hides an old-model Vauxhall Cavalier which relies for its acceptance on a tempting buying package as much as outright talent. In fact, it's not a bad effort, especially at the price. Apart from the Eighties-style facia, some tacky switchgear and cheap-looking cloth trim, the Espero seems well built. It boasts a strong complement of standard goodies too, including airbag, anti-lock brakes and air conditioning. Engines are 1.5, 1.8 or 2.0 petrol units.

A key strength of the Espero is the amount of space, both for passengers and luggage - it's up there with the best family saloons, with high levels of seat comfort for those in the back, even if the front seats are only average. On the road the Espero rides well enough with the bigger engines providing more than adequate performance. The power steering and slick controls make it an effortless drive. That it handles and steers with confidence, if not finesse, is quite acceptable for all but the more performance-oriented drivers.

Best All-Rounder: Espero 1.8 CDi

BODY STYLES:	Saloon	PRICE FROM:	£11,500
ENGINE CAPACITY:	1.5, 1.8, 2.0	MANUFACTURED IN:	Korea

Daihatsu HiJet

With the multi-purpose vehicle concept taking off in all directions it's hardly surprising that those manufacturers left without a contender have been scratching around to fill the void in their range. Daihatsu has come up with the Hijet, based on its pint-sized van with a few windows and three rows of seats added. Because it is so small, Daihatsu is hailing it as a new breed of people carrier. In truth, the whole plot just doesn't gel. What you get, of course, is a vehicle that drives and rides like a van and with all the refinement of, well, a van.

For all that, there's no denying that the Hijet does provide a lot of accommodation for not all that much money. But that's all it offers. You wouldn't really want to live with a car offering a ride this choppy, with sluggish acceleration, a slow cruising speed, nowhere to put the shopping - unless it's on a seat - and with the road manners of a shopping trolley. Anyone expecting a cut-price Renault Espace will be sorely disappointed, even with the recently announced 'luxury' SE model.

Best All-Rounder: Hijet

BODY STYLES:	Multi Purpose Vehicle	PRICE FROM:	£8,000
ENGINE CAPACITY:	1.0	MANUFACTURED IN:	Japan, Italy

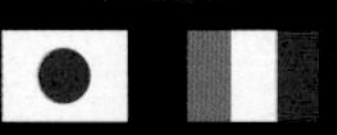

Daihatsu Charade

A facelift has brightened up the Charade, a new bonnet and headlamps giving this supermini a more charismatic face than in the past. At the same time the range has been confined to 1.3-litre five-door hatchbacks and a 1.5 saloon, all with power steering but with the airbag and sunroof still reserved for the option list. A cheaper three-door, with a lower specification, is on the cards.

In 1.3 guise, the Charade provides lively performance and it is now more refined than in the past, with a marked reduction in noise levels. Good road manners and a fine basic driving position make it easy to live with, the power steering a welcome feature. Oddly, the Charade offers a better deal for those in the back, with a high, comfortable seat and plenty of headroom, even though legroom is naturally tight. But the seat coverings are garish and in the front there just isn't enough lumbar support for long-term comfort.

Best All-Rounder: Charade 1.3 LXi

BODY STYLES: Hatchback, Saloon
ENGINE CAPACITY: 1.3, 1.5, 1.6
PRICE FROM: £7,000
MANUFACTURED IN: Japan

Daihatsu Fourtrak

There's a certain tractor-like charm about the Fourtrak that sets it apart from the other off-roaders. It looks tough enough to cope with any off-road or towing situation and that impression is reinforced once you look beneath and see the massive girder-like chassis and hefty suspension. Inside too, the overall effect of the hard-plastic facia and workmanlike switchgear is more of a working vehicle than a car, even though the top models come well equipped. All have two doors plus a side-hinged tailgate, but additional fold-down seats are fitted to some models to allow seating for seven.

The powerful 2.8 turbo-diesel engine endows the Fourtrak with impressive mid-range pull but has a harsh and hammery note. This lack of finesse is reflected in the stiff suspension which limits roll under cornering at the expense of a ride quality that at worst is appallingly choppy. This is a tough vehicle but it's not a particularly pleasant one to drive on the road.

Best All-Rounder: Fourtrak 2.8TDX

BODY STYLES: Estate
ENGINE CAPACITY: 2.8TD
PRICE FROM: £14,000
MANUFACTURED IN: Japan

Daihatsu Sportrak

The 'leisure off-roader' market has been growing rapidly in recent years as makers accept that most buyers, while desiring 4x4 ability, spend most of their time on solid roads. The Sportrak was one of the first of the breed, but now it's in its last throws before a replacement arrives in the spring of 1997. It's still capable of serious off-road work with its separate chassis construction, generous ground clearance and selectable but basic four-wheel drive. The two-doors with side-hinged tailgate arrangement make it work best as a two-person vehicle, but there is room for two more in the rear seats, although there's not a lot of room behind for luggage.

In the light of newer and more sophisticated compact 4x4s on the market, the Sportrak is beginning to appear markedly short on talent. A choppy ride and precarious handling do it no favours and although the 1.6 engine makes a fair effort at pushing it along, more power wouldn't go amiss. Off road the Sportrak is fun and capable: on road it has too many compromises.

Best All-Rounder: 1.6 EXi

BODY STYLES: Estate
ENGINE CAPACITY: 1.6
PRICE FROM: £10,000
MANUFACTURED IN: Japan

Dodge Intrepid

The Intrepid was the first of Chrysler's so-called 'cab-forward' cars, with the wheels pushed out to the corners to give cavernous levels of interior space. But this Dodge is a big car that's also fun to drive. Base models come with a 161 bhp 3.3-litre V6, with the more upscale ES version having a punchy 3.5-litre, 214-horse V6 driving the front wheels. It's a sporty-looker too, with its raked-back windscreen, ground-hugging nose and tall 16 inch wheels.

The driving fun comes from the Intrepid's taut, well-damped suspension, sharp steering and lusty, free-revving 3.5 V6. Couple these with a quick-reacting automatic and 225/60-section rubber and this Dodge stands out among its rivals. In keeping with its sporty image, it makes plenty of noise - engine roar and tyre rumble being the most intrusive offenders. The plastic interior trim needs to be more durable.

Best All-Rounder: Intrepid ES

BODY STYLES: Saloon
ENGINE CAPACITY: 3.3V6
PRICE FROM: $18,500
MANUFACTURED IN: US

Dodge Viper

If the original Viper didn't offer enough in the way of over-the-top aggression and sheer brutality with its in-your-face styling and 8.0-litre V10 truck engine, Chrysler has gone one step further. The new GTS is a fixed-head coupe which mimics the style of the old Cobra racers. It's lighter than the roadster, while both cars get much stiffer chassis to tighten up the handling. Work on the engine has both lightened it and increased the power by 50 bhp to a staggering 450 bhp.

It all makes for a brawny, he-man of a sports car which has none of the finesse found in a European supercar. Out of town the gut-wrenching power means it's hardy necessary to change gear, but the Viper's ample girth can be an embarrassment through narrow streets or country lanes. Inside it is all a bit too unsophisticated, with a hard plastic facia and, in the roadster, clip-on side screens and crude hood.

Best All-Rounder: Viper GTS

BODY STYLES: Convertible, Coupe
ENGINE CAPACITY: 8.0V10
PRICE FROM: £62,300
MANUFACTURED IN: US

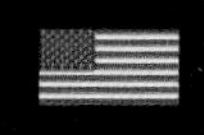

Donkervoort D8

Bearing an uncanny resemblance to the Lotus and Caterham Seven sports cars, to which the Dutch manufacturer owes a great deal of its heritage, the Donkervoort has been developed into a car sufficiently different to stand up on its own. Donkervoort has gone high-tech with the D8 - desperately wide wheels and tyres, racing car instrumentation and wafer-thin race seats give it the look of a car of the 90s, though inevitably it loses some of the purity of form which remains so vital to Caterham.

The structure is based around a steel space-frame chassis clothed in aluminium panels. Engines range from the Ford Zetec 16-valve producing 140 bhp to a 2.0-litre Escort Cosworth power unit, give a staggering turbocharged 220 bhp, or even 280 bhp if you are both rich and brave enough. If this isn't enough, the D8 Sport has been lightened by 10% with Kevlar panels, lightweight seats and the removal of most trim.

Best All-Rounder: D8 Zetec

BODY STYLES: Convertible
ENGINE CAPACITY: 1.8, 2.0
PRICE FROM: approx. £20,000
MANUFACTURED IN: Netherlands

Ferrari F355

The true driving enthusiast always buys the smaller Ferraris, for these are the most nimble and pleasing of all to drive quickly. The F355 is the best for years, certainly a big step over the 348 it replaced. First, the F355 presents a huge hike in power. With five valves per cylinder on the 3.5 litre V8, 380 bhp has been extracted, making this the most powerful non-turbo-charged car per litre of capacity you can buy. Then there is a Formula One-style undertray to help smooth the air flow and electronic control of the suspension.

The result is a car which not only comprehensively out-performs the old 348, in most circumstances it will out-accelerate the much bigger, more powerful Ferrari 512M. Acceleration to 60 mph is under 5 seconds, top speed 185 mph. As with all Ferraris, it as much the charismatic sound of the engine as its sheer performance which entrances drivers and passengers alike. Drivers have fun in other ways too, with a sweet gearchange and beautifully finished interior. The only real fault is the badly offset pedals.

Best All-Rounder: F355 Berlinetta

BODY STYLES: Coupe, Convertible, Targa
ENGINE CAPACITY: 3.5V8
PRICE FROM: £93,000
MANUFACTURED IN: Italy

Ferrari 456GT

Arguably the most beautiful Ferrari in years, the Pininfarina-designed 456 GT takes styling themes from great Ferraris of the past, notably the classic Daytona. Pitched as a Grand Tourer rather than an out-and-out supercar, the emphasis is therefore a little more on practicality - there's room for a couple of adults to squeeze in the back and the boot is a reasonable size, helped by the five tailored suitcases that come with the car. The interior is luxurious with Connolly leather everywhere.

Grand Touring Ferraris have always been front engined, and this is no exception. It's a mighty impressive engine too - a 5,437 cc 48-valve V12 which produces 442 bhp. That's enough to give a top speed of 191 mph and acceleration from rest to 60 mph in a mere 5.2 seconds. An automatic transmission 456 has recently been introduced; surprisingly there are plenty of Ferrari owners who don't want to change gear manually.

Best All-Rounder: 456 GT

BODY STYLES: Coupe
ENGINE CAPACITY: 5.5V12
PRICE FROM: £161,000
MANUFACTURED IN: Italy

Ferrari 550M

The replacement for the 512M takes on the mantle of Ferrari's top production two-seater from a new angle. Out goes the horizontally-opposed 12 cylinder engine, mounted in the middle of the car to give the best possible balance. In comes a front-engined V12, designed as much with practicality and comfort in kind as outright performance. Of course, the new 550M is still desperately fast, quicker in fact that 512M on road or track.

There's a six-speed gearbox which will be offered with steering wheel 'paddle' control in 1997, rather like the grand prix cars. Power is 485 bhp, the engine a development of that in the Ferrari 456. Everything has been focussed on ensuring this latest two-seater Ferrari is not only a stunningly quick car but also a practical one - something Porsche owners are used to, but not those owning Italian supercars. But whether the taller, less extreme design will turn out to be a classic is rather more open to question.

Best All-Rounder: 550M

BODY STYLES: Coupe
ENGINE CAPACITY: 5.5V12
PRICE FROM: approx. £145,000
MANUFACTURED IN: Italy

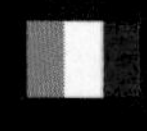

Ferrari F50

The most powerful and fastest road-going Ferrari in history sold out as soon as it was announced, but cars are still being built for delivery in 1997. The F50 is as close to a Formula One car as you can hope to get, with an engine genuinely derived from the Grand Prix car, mounted in a high-tech carbon-fibre monocoque chassis. The engine is based around that in Alain Prost's 1990 F1 Ferrari, enlarged to 4.7 litres to produce a whopping 520 bhp - enough to take the F50 to 60 mph in 3.7 seconds. The top speed of 202 mph, however, is well short of the McLaren F1's 230 mph, although the F50 costs only half the price.

Inside the F50 is racing car bare, with no sign of carpet and minimal fancy trim. It's unlikely that quietness and a soft ride are top of the list of attributes either, for the F50 is a car for a high-speed blast on a sunny day, no more no less. Delightfully impractical, the F50 is a Lotus Elise for the super-rich.

Best All-Rounder: F50

BODY STYLES: Convertible
ENGINE CAPACITY: 4.7V12
PRICE FROM: £350,000
MANUFACTURED IN: Italy

Fiat Cinquento

Fiat is the master, in Europe at least, of producing cars at the small, budget end of the spectrum. The Cinquecento is the latest in a long line of Fiat 500s, superseding the unmissed Fiat 126. Unlike the older rear-engined models, the Cinquecento is front-wheel drive and cleverly designed to provide room for four adults as well as a fair-sized boot. S and SX models come with a 900cc four-cylinder engine, the Sporting is 1.1 litres with a five-speed gearbox.

While no one will pretend the Cinquecento is a car for travelling long distances, for short trips it makes a lot of sense. It's compact and easy to park, comfort in the front is reasonable and there is loads of oddments space inside. Unlike previous Fiat 500s, even in its most basic form the Cinquecento offers acceptable levels of performance, while the more powerful Sporting version is a hoot to drive, a real Mini Cooper of the 1990s. All versions, however, have poor pedals and weak ventilation.

Best All-Rounder: Cinq' Sporting

BODY STYLES: Hatchback
ENGINE CAPACITY: 0.9, 1.1
PRICE FROM: £5,600
MANUFACTURED IN: Poland

Fiat Palio

The term 'world car' has been bandied around so many times the concept has become devalued. The Palio is different, developed by Fiat in Italy solely for developing markets worldwide. Initial production has already started in Brazil, with Argentina, Poland, Turkey and India to follow, and with plans in place for China, Chile, Egypt, Morocco, South Africa and Venezuela. The strategy is to be building a million Palios a year by 2000, which should make it the world's best seller.

Similar in size to the Fiat Punto, the Palio comes is six different body styles, three and five-door hatchbacks, saloon, estate, van and pick-up. Mechanically, it is simpler than cars for Western Europe, with fewer parts, a higher ground clearance to cope with worse roads and a design which is pitched towards more labour intensive factories. Engines will be familiar though, ranging from 1.0 to 1.6 litre, plus at least one diesel.

Best All-Rounder: Too soon to say

BODY STYLES: Hatchback, Saloon, Estate
ENGINE CAPACITY: 0.75, 1.0, 1.1
PRICE FROM: n/a
MANUFACTURED IN: Worldwide

Fiat Panda

Fiat has dropped the Panda from the UK market but it is still manufactured in Poland for sale to other European countries. A back-to-basics alternative to the plusher superminis, the Panda is available with a surprisingly broad mechanical package, which includes the option of four-wheel-drive and Selecta automatic transmission. Engines range from 750cc to 1.1 litres, and there has been a diesel in the past.

The Panda still has some strong points, not least the amount of space inside for a car of its price and the impressive amount of storage space for oddments. Initially conceived as a very simple car, with hammock-type chairs and little in the way of trim, it has been upgraded over the years, although little was achieved with its hard ride. A Panda is very cheap to run, but has fallen well behind on the safety features.

Best All-Rounder: Panda 1.0

BODY STYLES: Hatchback
ENGINE CAPACITY: 0.75, 1.0
PRICE FROM: approx. £5,500
MANUFACTURED IN: Italy

Fiat Punto

With the award for 'European Car of the Year' in 1995, Punto sales have gone from strength to strength - it's one of Europe's top selling superminis. With it's distinctive design, including huge rear tail lights and high waistline, there is no mistaking the Punto for any other supermini. All the usual options are there, including a vast range of petrol and diesel engines, three or five-door hatchbacks and a pretty cabriolet with a power hood.

Puntos are more entertaining to drive than most small cars. Even the 1.1-litre will zip through the gears with gusto, while at the other extreme the 1.4 Turbo has astonishing performance. The Punto Sporting combines the looks of the GT with a more insurable 1.6-litre engine. Inside there's little in its class to match the space on offer, with comfort for four adults. The facia design and general feeling of solidity is streaks ahead of earlier small Fiats. The ride is a bit firm and the steering heavy for parking, but that's about it for criticism.

Best All-Rounder: Punto 75 SX

BODY STYLES: Hatchback, Convertible
ENGINE CAPACITY: 1.1, 1.2, 1.4 Turbo, 1.6, 1.7D, 1.7TD
PRICE FROM: £6,900
MANUFACTURED IN: Italy

Fiat Bravo/Brava

Fiat's 1996 'European Car of the Year' offers buyers a distinctive alternative in the small family market. The Bravo is the three-door, with truncated tail, unusual oval rear light cluster and sporty appeal. The Brava is the same as far as the windscreen, but then has more conservative lines with the promise of more room for rear occupants and their luggage. The style continues inside, with a egg-shaped central facia unit which houses the built-in stereo and heater controls - always a talking point with those new to the car.

The broad range of engines runs from 1.4 to a five-cylinder 2.0-litre as well as diesels. The 1.4 goes well only when driven hard, whereas the 1.6 provides more relaxed, even sporting all-round performance. All are easy to drive, and most are fun, although the standard power steering is rather too light. The front seats are comfortable as long as adjustable lumbar support is fitted, but rear headroom is poor on Bravo/Bravas fitted with a sunroof.

Best All-Rounder: Brava 1.6 SX

BODY STYLES: Hatchback
ENGINE CAPACITY: 1.4, 1.6, 1.8, 2.0, 1.9D
PRICE FROM: £9,800
MANUFACTURED IN: Italy

Fiat Marea

Fiat's replacement for the Tempra, the Marea, went on sale throughout Europe in the autumn of 1996, and reaches the UK early in 1997. Derived from the Brava and Bravo hatchbacks, the Marea manages to be both longer and more elegant than either, and should offer significant improvements to the limited headroom for those in the back. The Weekend's estate-car flexibility is helped by a divided tailgate, just like the Tempra.

The increased size pushes the Marea into a class above the Brava, or so Fiat hopes, to compete with the likes of the Mondeo. So it offers Mondeo-sized engines, starting with a 1.6, through to a 113 bhp 1.8, with a 147 bhp 2.0-litre five-cylinder engine topping off the range. Diesel enthusiasts will not be disappointed either, with a choice of three turbo engines, including a 2.5-litre five-cylinder model. Early signs are that the Marea has even better refinement than the Brava.

Best All-Rounder: Too soon to say

BODY STYLES: Saloon, Estate
ENGINE CAPACITY: 1.4, 1.6, 1.8, 2.0, 1.9TD, 2.4TD
PRICE FROM: n/a
MANUFACTURED IN: Italy

Fiat Coupe

The breathtaking styling of Fiat's Coupe stands out from the conservatism of the Calibra and Probe. From its bulbous headlamps, Ferrari-like grille and clam-shell bonnet to the four round tail lights, the Fiat Coupe is uncompromising, with a style which is followed through to the interior where a body-coloured steel strip runs around the dashboard and doors. Power is provided by a 2.0-litre or a 2.0-litre turbo, but for 1997 these four cylinder units will be superseded by five-cylinder engines of similar capacity.

The Coupe is as stirring to drive as it is to look at. The original engines make a pleasing bark, and although the 2.0-litre is pleasing, the Turbo provides addictive levels of thrust, with the front wheels kept in check with a viscous-coupled limited slip differential. Grip and handling are superb. The Coupe is comfortable and rides well enough; adults will - just - fit in the back seat. The only serious problem is that this is not a hatchback.

Best All-Rounder: Coupe Turbo

BODY STYLES: Coupe
ENGINE CAPACITY: 2.0, 2.0Turbo
PRICE FROM: £17,900
MANUFACTURED IN: Italy

Fiat barchetta

Fiat's answer to the Mazda MX-5 is the highly stylised barchetta - with a lower case 'b'. The starting point is a shortened Fiat Punto platform, which means a transverse front engine and front-wheel-drive. This is not the usual sports car layout, although as Lotus used it to such good effect in the Elan, Fiat is on solid ground. Power is from an all-new 1.8-litre 16-valve variable valve timing engine, producing 130 bhp.

The barchetta is packed with pleasing styling features, from the chrome door handles which pop out at the press of a button, to the old-fashioned flaps for the fresh air vents to the hood which folds away beneath a plastic cover to give clean lines with the roof down. There are no complaints about the performance either, but the barchetta falls down in too many other areas - it drives more like a hatchback than a sports car, the ride and engine noise are unpleasant, the build quality isn't up to scratch and it's only available in left-hand-drive.

Best All-Rounder: barchetta 1.8

BODY STYLES: Convertible
ENGINE CAPACITY: 1.8
PRICE FROM: £14,300
MANUFACTURED IN: Italy

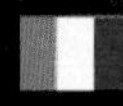

Ford Ka

With superminis of the Fiesta class becoming larger and larger, an opportunity has opened up for a whole range of smaller cars. Fiat has been eminently successful here, with the 126, Panda and Cinquecento. But Renault set the ball rolling in terms of styling with its Twingo, and now Ford follows with the Ka. Shown as a concept in 1994, the production version is closely similar, with curvaceous styling inside and out. Significantly, like the Twingo, the Ka is offered in just one body style with a single engine and two trim levels.

Aimed very much at the private buyer, the Ka is a three-door hatchback with a 1.3-litre engine and five-speed gearbox. If this sounds familiar it is, for the Ka makes use of much of the structure and mechanics of the Fiesta. The standard model gets an airbag, immobiliser and radio, while Ka 2 adds power steering, central locking and power windows. If the price is right, the Ka's character, distinctive styling and likely refinement are sure to make it a winner.

Best All-Rounder: Ka

BODY STYLES: Hatchback
ENGINE CAPACITY: 1.3
PRICE FROM: n/a
MANUFACTURED IN: Spain

Ford Fiesta/ **Mazda** 121

Ford's supermini was relaunched at the end of 1995 with a whole raft of changes that effectively overshadowed the fact that this is a development of the old Fiesta, still sold as the Classic. The obvious changes are a fresh nose, tailgate and lights, and a totally new dashboard which gives the Fiesta a classy appeal rarely found in a small car. Underneath much work has been done on the suspension, and there are two new 16-valve engines, a 1.25 and 1.4, as well as a reworked version of the 1.3 and diesel. The Fiesta is also sold by Mazda dealers with minor changes as the 121.

The result is a car which stands at the head of its class. The 16-valve engines are delightful, smooth and powerful. With power steering the Fiesta is a peach to drive, with an easy gearchange and excellent driving position. The ride and seat comfort for those in the front are about as good as you might expect. The downside is that the rear room is poor.

Best All-Rounder: Fiesta 1.25 LX

BODY STYLES: Hatchback
ENGINE CAPACITY: 1.25, 1.3, 1.4, 1.8D
PRICE FROM: £7,600
MANUFACTURED IN: UK

Ford Escort

The current Escort has undergone two facelifts since its introduction in 1990. Sales have always been sky high, with the Escort often topping the best-seller charts in the UK. Part of the success is down to the huge range available. Ford offers the Escort as a three and five-door hatchback, saloon, estate and convertible, as well as a van and the curious Courier, the van turned into a tall estate car. Engines range from the indifferent 1.3, 1.4 and diesels, to the much better 1.6 and 1.8 16-valve units. The impressive sports models, the RS 2000 and the RS Cosworth, both ceased production in 1996.

In its latest guise the Escort deserves its sales success. The interior is particularly successful, with a good feeling of quality and comfort, and room for four adults. The Escort is now much more satisfying to drive, with power steering standard on LX models upwards, even if the performance of most models is never going to set the world alight.

Best All-Rounder: Escort 1.6 LX

BODY STYLES: Hatchback, Saloon, Estate, Convertible
ENGINE CAPACITY: Petrol: 1.3, 1.6, 2.0i. Diesel: 1.8 turbo
PRICE: £10,000
MANUFACTURED IN: UK/ Germany/ Spain/ Poland/ Turkey

Ford Mondeo

The Mondeo receives a major update for the 1997 model year. A re-profiled nose with swooping headlamps and prominent grille, plus new bumpers and tail lights help freshen up the appearance. The engines remain largely the same, but there are improvements to the gearchange, suspension and rear kneeroom that should help keep the Mondeo in touch with more recent offerings like the Peugeot 406.

Not that there was much wrong with the original Mondeo - it was well thought out and good to drive. The interior design is particularly appealing, a well-designed facia, good seats and trim making the Mondeo a cut above many competitors. Three body styles are available, hatchback, saloon and estate, all with split folding rear seats. There is a wide range of good petrol engine, including a marvellously smooth 2.5-litre V6, plus a 1.8 turbo-diesel - the weak link. The Mondeo drives better than any previous family Ford, yet it manages to provide a comfortable ride too.

Best All-Rounder: Mondeo 1.8 LX

BODY STYLES: Hatchback, Saloon, Estate
ENGINE CAPACITY: 1.6, 1.8, 2.0, 2.5V6, 1.8TD
PRICE FROM: £12,800
MANUFACTURED IN: Belgium

Ford Probe

Ford took its time to respond to the success of Vauxhall's Calibra, eventually deciding to import the Probe from the States to fill the obvious hole in its model range. There are two versions available, a 2.0-litre 16-valve or 2.5 litre V6. While the smaller engine provides adequate but far from scintillating performance, the silky smooth V6 is a far more enticing prospect. It's from a Mazda, and indeed the underpinnings of the Probe are in reality those of the Mazda MX-6 coupe, although none the worse for that.

The Probe is good to drive, with better steering and handling than the Calibra, and an appealing interior which, unlike the Vauxhall, does not show its family car origins. Where it falls down, however, is in the amount of interior space. Headroom is limited for tall drivers, the rear seats are realistically suitable only for children, while the boot is shallow with a desperately high sill.

Best All-Rounder: 2.5 V6

BODY STYLES: Coupe
ENGINE CAPACITY: 2.0, 2.5V6
PRICE FROM: £17,800
MANUFACTURED IN: US

Ford Scorpio

The current Scorpio caused a stir when it was launched in late 1994. Bug-eyed headlamps and strip tail lights ensured that it stood out in a crowd, although some didn't like the reason why. Underneath the surface it's much the same as the old Granada, with rear-wheel-drive and a saloon or estate body, although the hatchback has now gone. The engine range has been revised for 1997, with a new 147 bhp 2.3 16-valve replacing both the 2.0-16v and the 2.9-8v, leaving the 2.0-8v and 2.9-24v V6 plus the 2.5 turbo-diesel.

Inside the Scorpio remains a hugely spacious car, with much mock wood which does surprisingly well in giving the car classier feel. Suspension developments result in much more control on winding roads at the expense of a firmer ride. The 2.0-litre 16-valve engine is a bit gruff sounding but provides handy levels of performance, although in the ultimate 24-valve form, the V6 Scorpio is a very refined and very quick car.

Best All-Rounder: Scorpio 2.3i Ghia

BODY STYLES: Saloon, Estate
ENGINE CAPACITY: 2.0, 2.3, 2.9V6, 2.5V6, 2.5TD
PRICE FROM: £18,700
MANUFACTURED IN: Germany

Ford Galaxy/**VW** Sharan **SEAT** Alhambra

This cooperative venture has been a huge success in its first year. Now that Seat has come onto the scene with its version - fitted with the desirable air conditioning as standard in the UK - the dominance may continue. Though there are differences between the Galaxy, Alhambra and Sharan they are in the most part subtle - different seat coverings and parts of external trim. Power choice is from a 1.9-litre direct injection diesel, 2.0 petrol or 2.8-litre VR6, all from Volkswagen, although the Galaxy uses Ford's own 2.0-litre engine, soon to be replaced by a new 2.3.

To date this is easily the most enjoyable people carrier to drive, with normal car-like controls and first-rate steering and handling. The VR6 gives wonderful performance but guzzles petrol, whereas Ford's 2.0 is a reasonable compromise. The fit and finish is extremely good, with much pleasing detail design. The seats are comfortable, though firm, with restricted foot room in the back row. Four-wheel drive arrives in '97.

Best All-Rounder: Galaxy 2.0 GLX

BODY STYLES: Multi Purpose Vehicle
ENGINE CAPACITY: 2.0, 2.8V6, 1.9TD
PRICE FROM: £16,700
MANUFACTURED IN: Portugal

Ford Escort (US)

Ford may be into building global cars these days, but this fresh-for-'97 Escort is for American consumption only. Developed from the old generation Escort, which had been around longer than George Burns, this curvy, slant-eyed newcomer has been transformed into a fun-handling, well-built commuting machine. Great value too; despite being completely redesigned, the new Escort keeps the old car's bargain-basement price tag.

Just one engine is offered with the '97 Escort; the familiar overhead-valve, in-line four which gets bumped-up in capacity from 1.9 to 2.0-litres. Power rises too, from 88 bhp to a more respectable 110 bhp. But it's the way the Escort handles itself on the road that impresses the most. Communicative steering, with oodles of front-end grip, give the car a nimble, sporty feel. In addition to the four-door, there's a roomy estate with a twin cam coupe due next year.

Best All-Rounder: Escort LX

BODY STYLES: Hatchback, Estate
ENGINE CAPACITY: 2.0
PRICE FROM: $11,500
MANUFACTURED IN: US

Ford Contour

Americans like their legroom. That's why Ford's US version of the European Mondeo - the Contour, along with its Mercury sister, the Mystique, received such a lukewarm reception when they debuted two years ago. Now the backs of the front seats have been scooped out to give an extra inch of knee room, and for '97 the rear seat is re-angled to provide a tad more space for legs and heads. Ford is hoping it will do the trick.

No one, however, has been moaning about the way this rounded-edged Ford, with its sloping rear screen, handles itself on the road. Impressive grip, with low-roll cornering, gives the car a strong, sporty feel. Like its Euro cousin, the Contour is powered by the 125 bhp 2.0-litre Zetec four-cylinder, with the smooth-spinning 170 bhp 2.5 V6 as an option. Next year, Ford's SVT performance division will offer a hotter, 200 bhp version with even sportier handling.

Best All-Rounder: Contour LX 2.0

BODY STYLES: Saloon
ENGINE CAPACITY: 2.0, 2.5V6
PRICE FROM: $14,000
MANUFACTURED IN: US

Ford Mustang

With a 4.6-litre V8 under its bonnet, in place of the venerable old Five-O, the Mustang is now more refined than it's ever been in its 32-year history. The new engine produces 215 bhp - the same output as the previous five-litre - but it's much smoother and less raucous from idle all the way to its 6000 rpm red line. Want something even badder? Try the 305 bhp 32-valve Mustang Cobra, available in coupe or drophead form. Want cheaper? Then there's the 150 bhp 3.8-litre V6. Each one is fun to drive and very competitively-priced.

With a new, high-tech engine driving the Mustang's rear wheels, you might have expected the new 4.6 to be quicker than its predecessor. Not so. Acceleration is the same with 0-60 mph sprinting taking around 6.8 secs.

The Cobra is a different story; 60 mph appears in 5.7 secs. All Mustangs are a blast to drive in a crude, vintage kind of way. While there's no shortage of grip, the car does get thrown off line by bumps. Still a great-looker though.

Best All-Rounder: Mustang GT 4.6

BODY STYLES: Coupe, Convertible
ENGINE CAPACITY: 3.8V6, 4.6V8
PRICE FROM: $15,500
MANUFACTURED IN: US

Ford Taurus

That quirky, head-turning, ovoid styling hasn't prevented Ford's new Taurus from becoming America's best-selling car. Indeed, its popularity continues to grow as Ford re-aligns its previously over-ambitious pricing. Almost the same length as an S-class Merc, the Taurus offers seating for six and plenty of space for luggage. The standard engine is a 3.0-litre, 145 bhp V6, though the optional 200 bhp 3.0-litre Duratec V6 provides much smoother motoring. In addition to the saloon, there's an even more striking estate version.

To spice-up the Taurus image, Ford has just put a more-pokey V8-engined version into its rounded four-door. Powered by a 235 bhp Yamaha-made 3.4, the new Taurus SHO combines sporty car performance with family car space at an affordable price. It's not perfect; it only comes with an automatic, for starters. And it's actually a tad slower than the previous V6-engined SHO. But it handles enthusiastically, and emits a deliciously muted V8 growl.

Best All-Rounder: Taurus LX

BODY STYLES: Saloon, Estate
ENGINE CAPACITY: 3.0V6, 3.0-24v V6
PRICE: $18,500
MANUFACTURED IN: US

Ford Expedition

When a Mountaineer just ain't big enough, Ford has a new Expedition to take-up even more space in your driveway. New for '97, this XXL sized, nine-seat off-roader is based on Ford's rough and rugged F-150 pick-up truck platform. If that sounds a little too much on the utilitarian side, the Expedition does come with optional air suspension for a smoother ride. Power plants include the 215 bhp 4.6-litre V8, with the option of a 5.4-litre V8.

Ford's newly-minted F-150 truck isn't a bad place to start when producing a rugged, full-size 4x4. It has a strong, stiff platform, well-judged suspension and most un-truck-like power trains. Despite its bulk, the giant-sized Expedition - it's 19 inches longer than a Range Rover - is an impressive family hauler. Both V8s provide relaxed, refined performance, though getting this leviathan off the line takes time. Inside, it's as big as most living rooms and as nicely-furnished.

Best All-Rounder: Expedition 4.6

BODY STYLES: Estate
ENGINE CAPACITY: 4.0, 5.0V8
PRICE: $30,000
MANUFACTURED IN: US

Ford Falcon

Two domestic manufacturers vie for sales supremacy in Australia, Ford and Holden. Both produce big, comfortable saloons capable of travelling huge mileages in a land where petrol is cheap. Ford's offering is the Falcon, which seats five adults with ease in either saloon or wagon body styles. Australians don't mess around with small engines. The standard Falcon gets a 210 bhp 4.0-litre six-cylinder unit, with a 5.0-litre V8 as an alternative. The V8 comes only with automatic transmission.

The interior of the Falcon isn't particularly inspired - the base model is a cheap car compared with imports and the facia and seat coverings look it. The situation does improve further up the range, with Ford offering a whole family of Falcon-based derivatives - Futura, Fairmont, Fairlane and LTD. All provide high safety standards, including the option of an elongated front passenger airbag which fills the facia between the door and the driver's bag.

Best All-Rounder: Futura 4.0

BODY STYLES: Saloon, Estate
ENGINE CAPACITY: 4.0, 5.0V8
PRICE FROM: Aus $29,500
MANUFACTURED IN: Australia

Holden Commodore

For eight years the Australian manufacturer Holden was jointly owned by General Motors and Toyota. It built the Commodore, Nova (Toyota Corolla) and Apollo (Toyota Camry) locally and imported the Barina (Corsa) and Calibra from Europe, and a whole host of Isuzu-General Motors off-roaders from Japan. Then in mid-1996 the partnership was dissolved and in the long term Holden will concentrate just on the GM vehicles.

A locally-built Holden is Australia's most popular car and it has been improved and refined for 1997. The Series II Commodore has a more refined and more powerful V8 engine, a new manual gearbox, more supportive seats and a better interior. Other models are developed from it. The Calais and Statesman have a longer wheelbase, with different nose and rear window treatment. Engine choice is usually between a smooth 3.5-litre V6 or a 5.0-litre V8 for the sports models, the Commodore S and SS.

Best All-Rounder: Commodore V6

BODY STYLES: Saloon, Estate
ENGINE CAPACITY: 3.8V6, 5.0V8, 5.7V8
PRICE FROM: Aus $23,000
MANUFACTURED IN: Australia

Honda City

The City is Honda's answer to the increasing demand for cars from customers in non-Japanese Asian countries. Initially manufactured in Thailand, by the end of 1997 the City will be built and sold in Malaysia, Indonesia, Taiwan, the Philippines, Pakistan and India. Though it has some obvious similarities to the Civic, it is a brand new design which is different in many aspects under the skin. The suspension, for example, is of a simple strut design rather than the sophisticated multi-link type used in Europe, which helps keep the manufacturing costs in check.

The City is built only as a four-door saloon with a 1.8-litre four-cylinder engine. With a long wheelbase it is claimed to be one of the roomiest cars in its class. Owners won't necessarily get cloth upholstery, electric windows or power steering, but all models receive air conditioning and a stereo as standard.

Best All-Rounder: City LXi

BODY STYLES: Saloon
ENGINE CAPACITY: 1.3
PRICE FROM: n/a
MANUFACTURED IN: Thailand

Honda Civic

It may have Japanese parentage but Honda's Civic five-door and the forthcoming estate are British to the core and closely related to the Rover 400. Differences centre on the Honda engines of 1.4, 1.5 and 1.6-litres plus different front and rear styling and interior trim. Confusing the picture are the other Civics - three-door, saloon and coupe - which are half a generation ahead of the five-door and distinguished by their bug-eyed headlamps.

All Civics share a tightly assembled feel, a sporty character and are very easy to drive. Pick of the range has to be the five-door model with its fine road manners and the best ride of any small Honda yet. It also boasts some excellent safety features. The only negative point of any note is the mediocre rear passenger space, but some might cite the poor mid-range response and a vocal character to some of the engines as reasons to quibble. Otherwise this is one of the very best of the 'Escort' class cars.

Best All-Rounder: 1.5i VTEC-E 5-

BODY STYLES: Saloon, Hatchback, Estate, Coupe
ENGINE CAPACITY: 1.4, 1.5, 1.6, 1.6 VTi
PRICE FROM: £12,000
MANUFACTURED IN: UK/ Japan/ US

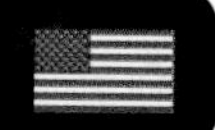

Honda Accord

Honda's Accord has been facelifted in the past year, with a fresh nose, revised seats and some new engines. The saloon is closely related to Rover's 600, and like the Rover it is built in the UK. Accord coupes and estates, on the other hand, hail from Honda's plant in the USA, and although they might look similar, are in fact of a different generation of car.

To live with, however, there's not much to choose. The new front seats provide good support, the chassis provides extremely satisfying handling and each engine in the range, including a Rover turbo-diesel, gives strong performance. Suspension changes have softened the ride so the latest cars are noticeably more comfortable. Hondas are well built, too, but the Accord is not without its weaknesses - rear comfort and space are poor for a car of this size and it still lacks the immediate classy appeal that works so well in the Rover.

Best All-Rounder: Accord 1.8i

BODY STYLES: Saloon, Estate, Coupe
ENGINE CAPACITY: 1.8, 2.0, 2.2, 2.7V6, 2.0TD
PRICE FROM: £14,800
MANUFACTURED IN: UK/ US

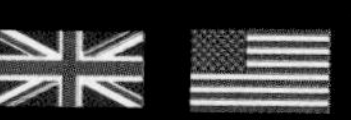

Honda Shuttle

Honda's Shuttle is pitched in at the top end of the rapidly expanding people carrier market, offering a degree of luxury that few others can match. The idea is that extra space and comfort can be offered by restricting the first two rows to four 'captain's chairs' to give a sort of elite travel that's one step better than an ordinary car. The third row of seats simply folds away into the floor when not needed, which boosts the otherwise restricted load space. It is a concept that hasn't been entirely convincing, however, and for 1997 Honda will offer a version with a full seven seats.

Not only is the Shuttle's design a good one, but it drives well too. The handling is sharper than anyone has a right to expect and the ride is supple. The 2.2-litre Accord engine would benefit from more power, especially as auto is standard, but it's civilised and cruises happily at speed. A comprehensive specification helps to justify the high pricing.

Best All-Rounder: Shuttle 2.2

BODY STYLES: Multi Purpose Vehicle
ENGINE CAPACITY: 2.2
PRICE FROM: £23,600
MANUFACTURED IN: US

Honda Prelude

Honda's Prelude has evolved over the years from a coupe offering more in the way of style than substance to a genuine sporting car that is a joy to drive. The downside is that the practical aspects have been pushed to the back burner, with rear seat space far too cramped for comfort.

Honda reckons its all-new 1997 model combines the best of both attributes. Both rear seat accommodation and luggage space are significantly increased and the new interior design brings improved ergonomics and seat comfort for the driver. Mechanically much will be familiar - the 2.0-litre 134 bhp and 2.2 VTEC 185 bhp engines and the all-round double wishbone suspension. But the new car also gets the option of electronic four-wheel steering on the VTEC Prelude, and an advanced four-speed gearbox from the NSX which allows the driver to choose clutchless manual changes or fully automatic operation.

Best All-Rounder: Prelude 2.2 VTEC

BODY STYLES: Coupe
ENGINE CAPACITY: 2.0, 2.2
PRICE FROM: approx £19,000
MANUFACTURED IN: Japan

Honda Legend

That Jaguar, Mercedes-Benz and BMW dominated the luxury car market with such assuredness for so long could only mean one thing - the Japanese car manufacturers would need to attack that area as successfully as they have all other aspects of the motor industry. But it hasn't been easy, with only Toyota making a convincing case for itself with its Lexus range of products. Now Honda is having another go with its new Legend, pitched at the cheapest Jaguar and BMW 7-Series saloons.

It certainly is beautifully appointed inside, full of soft leather, electrical adjustment for almost everything and limousine-like space. The V6 engine is now 3.5-litres and provides smooth, seamless performance - there's no doubt the Legend provides a very relaxed environment in which to travel. For the price this Honda offers a fine package, but has Honda managed to make its luxury car really desirable enough to be a success?

Best All-Rounder: Legend 3.5

BODY STYLES: Saloon
ENGINE CAPACITY: 3.5V6
PRICE FROM: £33,600
MANUFACTURED IN: Japan

Honda CR-V

Car manufacturers continue to search for undiscovered niches. Honda thinks it has found one and filled it with the new CR-V. Looking a little like Toyota's hugely successful fun car, the RAV4, but closer in size to the Land Rover Discovery, the CR-V aims to combine off-road ability with driving enjoyment, as well as offering the refinement of a passenger car. Power is provided by a 130 bhp 2.0-litre petrol engine. Permanent four-wheel-drive is standard, automatic transmission an option.

Unlike any other off-roader the CR-V has a completely flat floor, like an MPV. This allows for changeable seating positions and a central 'walk-through' from front to back. Under the rear floor there's a concealed luggage area, the lid of which can be removed and used as a table. With a horizontally-split tailgate and a power outlet in the luggage area, the CR-V looks designed to provide a perfect alternative to the race-course picnic vehicle.

Best All-Rounder: CR-V 2.0

BODY STYLES: Estate
ENGINE CAPACITY: 2.0
PRICE FROM: n/a
MANUFACTURED IN: Japan

Honda CRX

This diminutive two-seater sports car with its neat removable roof has acquired a cult-like status in some European countries. That's not the case in Britain, however, where it's seen as just a little too effete and too expensive. Some would regard the optional electric roof that stores itself neatly in the boot as pure novelty, but there's no denying that this is a thoroughly effective sports car beneath the toy-like veneer. That's certainly so with the discontinued VTi model and its 158bhp engine, but even the ESi has 123bhp at its disposal, which is enough to entertain.

It's fun to drive too, with accurate steering, plenty of grip and crisp and well-balanced handling. It certainly humbles the likes of the Vauxhall Tigra, but what is missing is the joie de vivre that any Mazda MX-5 driver appreciates. Excellent build integrity, a firm but supple ride and a pleasant, well-specified cabin makes this a civilised sporting car. It just needs a little more soul.

Best All-Rounder: CRX 1.6i ESi

BODY STYLES: Coupe
ENGINE CAPACITY: 1.6
PRICE FROM: £18,100
MANUFACTURED IN: Japan

Honda NSX

This high performance machine is as much a statement of Honda's engineering mastery as it is a car. Created to show the considerable bounds to which Honda's technology has extended, the five-year old NSX still has few peers for outright dynamic ability. Technology abounds, from the aluminium monocoque body and forged alloy suspension components to the 24-valve V6 engine with its variable valve timing and astronomically high rev range. As the Japanese interpretation of a Ferrari, it does a thorough job.

Few could fault the Honda for its dynamic prowess. Where the engine lacks outright power it makes up for it with a sweet gearchange and the spine-chilling wail it emits once beyond 5,000rpm. With a chassis of rare brilliance, it's quicker through S-bends than anyone has a right to expect, it steers with utter precision and its braking and grip are simply extraordinary. Faults are few, but some are disappointed by the plain interior.

Best All-Rounder: NSX Coupe

BODY STYLES: Coupe, Targa
ENGINE CAPACITY: 3.0V6
PRICE FROM: £69,500
MANUFACTURED IN: Japan

Hummer

Looking to start your own war? AM General has just the vehicle for you. The Hummer made a big name for itself fighting the good fight in Operation Desert Storm. Then Arnold Schwarzeneggar persuaded AM General to build a civilian version - you don't argue with Arnie - and the rest is history. Yes, it's crude, yes it's bigger than the average football stadium, and no, you'll never find a place to park it. But it will get you noticed.

The Hummer is the ultimate off-roader. Sixteen inches of ground clearance, a low centre of gravity, four-wheel drive and the unquestionable ability to climb up the outside of most tower blocks, make it unbeatable in the world's war zones. Or on the M25. And remember, no one will ever cut you up. Power comes from a GM-sourced 6.5-litre turbocharged V8, or a 5.7-litre V8 petrol engine with 300 lb ft of torque. Don't ask about fuel consumption.

Best All-Rounder: Hummer Diesel

BODY STYLES: Estate, Convertible
ENGINE CAPACITY: 5.7V8, 6.5V8D, 6.5V8TD
PRICE FROM: $43,000
MANUFACTURED IN: US

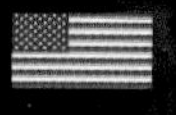

Hyundai Accent

The Accent is Hyundai's first completely 'home-grown' car, designed, developed and engineered in Korea. Two body styles are available - saloon and hatchback, though Hyundai optimistically calls the three-door hatchback the Coupe. Each has very distinctive, rounded styling that is unusual in this class of budget small family car. Engine choice is either a 1.3 or 1.5, with an automatic option on the bigger unit.

The Accent's key strength is its ease of driving. The quality of the power steering (not on all models), gearchange and clutch and the willing engines make this one of the simplest cars on the road to get to terms with. Ride and noise levels are restrained too. Getting into the Accent is a bit of a drop down, but once there the front seats are supportive and reasonably comfortable. Things are not so good in the back, where seat comfort is mediocre, while the whole interior is plain and characterless. But at the price, and with a great warranty, the Accent is excellent value.

Best All-Rounder: Accent 1.3 LSi

BODY STYLES: Coupe, Saloon
ENGINE CAPACITY: 1.3, 1.5
PRICE FROM: £7,000
MANUFACTURED IN: Korea

Hyundai Lantra

That Hyundai is getting serious about its export markets can be seen from the new Lantra. With a style which makes it stand out from the more conservative opposition, Hyundai hopes that the evident quality and keen pricing will take sales away from the likes of the Mondeo and Vectra. Two body styles are on offer, saloon and estate, with 16-valve engines of 1.6 and 1.8 litres.

It's arguable whether the Lantra really does compete head-on with the above mainstream cars - it's a touch too small, especially in terms of headroom. But otherwise it is an extremely competent family vehicle which contends well both in terms of comfort and driving ability. The 1.6 is lively and willing, and hardly any slower than the 1.8. There is superb speed-sensitive power steering which helps make the Lantra a pleasing driving experience. The ride and seats may be a bit firm, but the seats are well bucketed and provide plenty of support.

Best All-Rounder: Lantra 1.6 GLSi

BODY STYLES: Saloon, Estate
ENGINE CAPACITY: 1.6, 1.8
PRICE FROM: £10,000
MANUFACTURED IN: Korea

Hyundai Coupe

After slow beginnings, this Korean manufacturer has started to strike gold with its latest cars. The new Lantra-derived Coupe shows just how far the company has risen from its early days of cheap and not very cheerful small hatchbacks. It competes stylishly with the likes of the Calibra and Probe, offering 2+2 accommodation, a sporty interior and a new 2.0-litre engine. Pricing is very competitive, starting at under £15,000 for a version which includes practically everything you might choose, apart from air conditioning, leather interior and CD player, all of which come on the Coupe SE.

While the engine isn't especially powerful for a 2.0-litre, the Coupe turns out to be fun to drive, with sufficient power and a very well-developed chassis. The heavily bucketed front seats are very comfortable, but it's incredibly cramped in the rear, while the boot space is minimal too. And it's a pity about the compulsory sunroof - eats too much into the headroom.

Best All-Rounder: Coupe 2.0

BODY STYLES: Coupe
ENGINE CAPACITY: 2.0
PRICE FROM: £15,000
MANUFACTURED IN: Korea

Hyundai Sonata/ Marcia

With a new nose and tail treatment for 1997, this large saloon proves Hyundai really has come a long way over the last few years. Not only does the Sonata look convincing on first acquaintance, it delivers the goods in almost every area – except for those demanding a degree of sporting promise. For the driver who simply prefers to cruise, the Sonata could be a good choice with its generous specification and plenty of space within the comfortable cabin. It's good value too, and it boasts many of the latest safety features.

Performance from the 2.0-litre engine is reasonable, bearing in mind the car's size, but the 3.0 V6 with its auto box seems lost in a lethargic doze even when it's trying. Suspension improvements for '97 will hopefully improve the ride, but it remains to be seen what effect it will have on the steering, which was hardly what you would describe as sharp. Yet, driven in a more gentle fashion, few of these minor misdemeanours are ever exposed.

Best All-Rounder: Sonata 2.0 CD

BODY STYLES:	Saloon	PRICE FROM:	£13,400
ENGINE CAPACITY:	1.8, 2.0, 3.0V6	MANUFACTURED IN:	Korea

Hyundai Grandeur/ Dynasty

Back home in Korea Hyundai produces a full range of vehicles, topped off by the luxury Grandeur. Powered by a 3.0-litre V6 mated to an electronic four-speed automatic transmission, it is the biggest selling luxury car on its home market. Electronics feature big-time, controlling the gear changes, suspension settings, front and rear seats and the interior temperatures. A cool box is mounted in the rear shelf, chilled by the air conditioning. The hi-fi is a work of art, with balance and acoustics adjustable to suit the number of people in the car.

With so many gizmos, the cynical might expect the car itself to be something of an anti-climax. The V6 engine may produce a healthy 205 bhp but the indifferent performance - the top speed is merely 118 mph - is the result of the large amount of weight the Grandeur carries. There is also an upmarket version called the Dynasty, which is fundamentally the same car but with a nose and tail job and the option of a 3.5-litre V6.

Best All-Rounder: Grandeur 3.0 V6

BODY STYLES:	Saloon	PRICE FROM:	n/a
ENGINE CAPACITY:	2.0, 3.0V6	MANUFACTURED IN:	Korea

Jaguar/ Daimler XJ6

Many breathed a sigh of relief when Jaguar introduced its current range of saloons in 1994. Moving back to the round headlamps and flowing curves of past Jaguars there is no denying this is an exceptionally beautiful car. As before the engine choice is between a 3.2 or 4.0 six-cylinder, or 6.0-litre V12, but now there's additional excitement. The XJR, powered by a supercharged 4.0-litre, has the performance of an Aston Martin. If that's too expensive, the 'Sport' models have tighter suspension with the standard engines.

It goes almost without saying that the levels of comfort and refinement of the XJ saloon are world class. The interior is particularly appealing, with a well crafted mixture of leather, wood and stainless steel, though it's undoubtedly old-fashioned. Front and rear the ride, noise levels and comfort of the seats are praiseworthy, and although the room for those in the back seats isn't generous,the new long-wheelbase version helps.

Best All-Rounder: XJ6 4.0 Sport

BODY STYLES:	Saloon	PRICE FROM:	£30,000
ENGINE CAPACITY:	3.2, 4.0, 4.0 Supercharged, 6.0V12	MANUFACTURED IN:	UK

Jaguar/ Daimler XK8

Thirty-five years after the unveiling of Jaguar's beloved E-Type its spiritual successor was exhibited in Geneva. Replacing the XJS, the XK8 is less a grand touring coupe, instead harking back to the more sporting style of the E-Type. So the bodywork - coupe and convertible - has that characteristic feline look, its low nose accentuated by the oval grille, a high waist-line and muscular haunches over the rear wheels. Inside there's a classic cockpit theme with a pronounced dashboard cowl and traditional wood and leather.

Of course standards have moved on enormously, so the roof on the convertible will be power operated, while with Ford's input the quality and durability of the XK8 should be light years ahead of anything in the past. Beneath the surface, the structure of the XK8 owes a lot to the XJS. But the power unit is all-new, a 4.0-litre V8 engine coupled to Jaguar's first five-speed automatic gearbox - manual transmission will not be an option.

Best All-Rounder: XK8 convertible

BODY STYLES:	Coupe, Convertible	PRICE FROM:	approx. £45,000
ENGINE CAPACITY:	4.0V8	MANUFACTURED IN:	UK

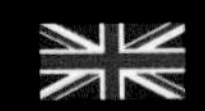

Kia Pride

For a company with such strong ambitions, it seems a little odd that Kia perseveres with the Pride. Discarded by Mazda in 1991, Kia took over manufacturing of the Japanese manufacturer's 121 model and have done very little to it since. The Korean-built hatchback is still notable for its low cost and ample interior space, but the supermini race has accelerated in recent years and the Kia Pride has been left high and dry. These days, safety, space, build quality and a reasonable specification are taken as read, but the Kia lags behind.

A Spartan specification and a dowdy interior with some cheap-looking plastics put the Pride at a disadvantage straight away, and the nose-heavy handling, the woolly steering and the bouncy ride make it seem all the more out of its depth. One saving grace is the engine; it delivers reasonable performance and excellent economy. For those who desire cheap and basic transport, it does the job, but there are better alternatives.

Best All-Rounder: Pride 1.3LX

BODY STYLES:	Hatchback	PRICE FROM:	£5,500
ENGINE CAPACITY:	1.1, 1.3	MANUFACTURED IN:	Korea

Kia Mentor

Kia, like so many Korean car-makers, is an ambitious company that successfully opens up new markets by offering good-value deals on well proven, Japanese-derived products. The Mentor is such a car. By using a combination of Mazda technology in a no-nonsense, well equipped package that contrives to do everything adequately well, the result is a car that has widespread appeal but no pretensions of class-leading ability. Initially available only as a saloon, 1997 sees a five-door hatchback join the range.

The styling is inoffensive but forgettable; so is the interior, but no less effective for that. The Mentor has adequate space for an Escort-sized car, it's well trimmed, and the equipment spec is far from mean unless safety is a priority. Driven gently, it will prove adequate to the task. Extend it further and the suspension becomes too lively, the 1.6-litre engine gets raucous and the front tyres begin to push wide through the turns.

Best All-Rounder: Mentor 1.6GLX

BODY STYLES:	Hatchback, Saloon	PRICE FROM:	£8,500
ENGINE CAPACITY:	1.6, 1.8	MANUFACTURED IN:	Korea

Kia Sportage

In the group of smaller off-roaders, none comes more stylish than the Kia Sportage. Think of it as a tall, smallish estate car with off-road ability and you'll get the right idea. It's a concept that works surprisingly well. Sparkling performance is on offer from the 2.0-litre engine, the handling is secure and as car-like as a tall 4x4 gets, and the ride is absorbent if a little bouncy. It's quiet and refined on the road and well mannered when driving on rough tracks. And if things get tough you simply snick it into four-wheel drive.

A high seating position means everyone has a good view, and there's room for five with far more space than you would get with a conventional car. Practicality is marred to some extent by the loading arrangement which means the spare wheel carrier must be swung aside before lifting the tailgate. Also, the rear seats occupy too much space when stowed. The Sportage is now built in Germany, with a shorter three-door version due soon.

Best All-Rounder: Sportage 2.0 GL

BODY STYLES: Estate
ENGINE CAPACITY: 2.0, 2.0D, 2.0TD
PRICE FROM: £13,300
MANUFACTURED IN: Korea/ Germany

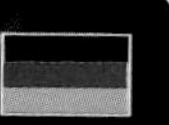

Kia Avella

Kia's aspirations for a larger slice of the world's car market draws a step closer with the Avella. This Escort-sized hatchback follows usual Kia practice in drawing heavily on borrowed technology. Beneath the skin it's essentially a current-model Mazda 323. So it gets a choice of Mazda-designed 1.3 and 1.5-litre engines and also inherits the 323's safety structure. It is also clear that the fastback styling is only very thinly disguised from the Mazda's. It's certainly distinctive but rear passenger space is on the tight side.

At the moment, this model is not available in Britain, although with the smaller Kia Pride looking increasingly uncompetitive among newer supermini designs, it seems likely we'll see the Avella before long. In 1.3 form the 75 bhp engine is particularly economical in conjunction with the low-drag body design, while the 1.5 produces more spirited performance. Kia's marketing strategy means the Avella's pricing will be aggressively competitive.

Best All-Rounder: Avella 1.5

BODY STYLES: Hatchback
ENGINE CAPACITY: 1.3, 1.5
PRICE FROM: n/a
MANUFACTURED IN: Korea

Lada Riva

Basic no-frills transport is the name of the game here. The Lada Riva is a past-generation car that's stuck in a time warp. Originally derived from the long-defunct Fiat 124, it's built in Russia and is possessed with all the sophistication of a house brick and the charm of a Soviet general. The 1.5-litre saloon gives adequate performance, despite having only four gears; an estate version with five gears shares the same engine.

The Riva might have a basic and Spartan specification but this does mean it's cheap to maintain. There's not much to go wrong and what there is can probably be fixed by the home mechanic. The plain interior is straight out of the Seventies with its nasty plastics, and there's little more space inside than a modern supermini. It rides well enough in a squashy manner but the handling and refinement are dire by today's standards. As ever, it relies solely on low price for survival.

Best All-Rounder: Riva 1.5E Estate

BODY STYLES: Saloon, Estate
ENGINE CAPACITY: 1.2, 1.3, 1.5
PRICE FROM: £4,500
MANUFACTURED IN: Russia

Lada Samara

Where rival Skoda - even before Volkswagen's involvement with the company - has managed to show that cars from eastern Europe don't have to be dismissed at first sight as crude and unpleasant, Lada has had much more difficulty. The Samara is due to be replaced in 1997 with the new 2110 range, and not before time. The current car has its roots firmly entrenched in the Seventies. The square-rigged shape and the nasty plastics within simply don't bear comparison with even the most basic of the new breed of supermini.

What it does have in its favour is price. This is one of the cheapest cars on sale, yet it's about the same size as an Astra in hatch form. There's also a saloon version, but it's no prettier. Power is quite acceptable in 1.5 form, barely so with the 1.3 and lethargic in 1.1 guise. None is particularly quiet, especially when worked hard, as each model needs to be. As expected, the chassis is soggy and ill-controlled; the equipment and safety spec minimal.

Best All-Rounder: Samara 1.3 S

BODY STYLES: Hatchback, Saloon
ENGINE CAPACITY: 1.1, 1.3, 1.5
PRICE FROM: £4,500
MANUFACTURED IN: Russia

Lada Niva

This rugged all-terrain vehicle is one of the cheapest on the market, so understandably it sports little in the way of sophistication or refinements. A basic selectable four-wheel drive system marks this out as a utility 4x4 rather than a leisure mud-plugger, although the third generation machines, introduced in 1996, received new seats and interior, improving the comfort and convenience from the early dire levels. Even so, originally conceived for the Russian army, the Lada Niva is a one off that will never be status symbol.

As a road car, the Niva remains pretty basic transport. Even with a new 1.7 litre engine with fuel injection, performance is modest in the extreme and there's no power steering to ease the high parking loads required. Inside, there's plenty of room for four hefty adults, but not much of a boot space behind the seats. Given a long enough straight, it will work its way noisily up to about 80mph, but show it a muddy track and the Niva feels far more confident.

Best All-Rounder: Niva Hussar

BODY STYLES: Estate
ENGINE CAPACITY: 1.6, 1.9D
PRICE FROM: £8,000
MANUFACTURED IN: Russia

Lamborghini Diablo

The original Diablo keeps spawning derivatives, and now the choice has risen to five. The standard mid-engined Diablo is powered by a 492 bhp, 5.3-litre V12, designed to ensure a top speed of 202 mph. The Diablo VT is a four-wheel-drive version. The Roadster, introduced in 1996 with a removable roof, uses the VT's four-wheel-drive transmission. Also new for '96 was the Diablo SV (Sport Veloce), the most driver focused of all the road cars. With more power, less weight and lower gearing, acceleration is even sharper, although top speed is down to a mere 186 mph. The SV-R is a racer.

In its stripped out SV form the Diablo is outrageously fast and outrageously noisy. But it's an intoxicating noise which gets the pulse racing ever before the clutch is engaged. The levels of grip are huge but the driver can adjust the setting of the anti-roll bar and the traction control from his seat to tweak things to perfection.

Best All-Rounder: Diablo SV

BODY STYLES: Coupe, Convertible
ENGINE CAPACITY: 5.7V12
PRICE FROM: £125,000
MANUFACTURED IN: Italy

Lancia Y

1996 saw the much-needed introduction of the all-new Y as the replacement for the Y10. This one immediately showed far more promise, taking as its starting point the chassis of the Fiat Punto, then adding the Punto's 1.2-litre or the Brava's 1.4-litre engine. The styling sets new standards for superminis, successfully giving this small car a sense of quality unavailable elsewhere. Amazingly the Y is available in 112 colours - buyers get a 1:25 scale model in their chosen shade then have three weeks to change their mind.

Strong styling themes continue inside the Y. The instruments and controls are located in the centre of the dashboard, leaving nothing but a shelf in front of the driver. The seat covering material is carried through to the facia surface, the combined effect being a remarkably classy interior. But to drive the Y doesn't match up to its looks. While the 1.4 engine is enjoyable, the 1.2 is too noisy and slow. And like the old Y10, the suspension is too firm.

Best All-Rounder: Y 1.4

BODY STYLES: Hatchback
ENGINE CAPACITY: 1.2, 1.4

PRICE FROM: approx £7,600
MANUFACTURED IN: Italy

Lancia Delta

In many ways Lancia's Delta is a casualty of Fiat-Group politics. Launched three years ago but never offered for sale in the UK, this Escort-sized hatchback was originally planned to reach production in the mid-80s. Until, that is, successive policy changes at Fiat stalled its introduction again and again. So the Delta is a product of a past generation, and it shows.

The five-door body style, reminiscent of the Alfa Sud, is the most attractive feature, but the space efficiency and the general dynamics are a level behind the expected standards today. The ride, for example, is turbulent at lower speeds, especially over urban streets, the noise refinement is poor, the facia looks dated and the handling of the mainstream models is decidedly dull with a nose-heavy bias. Engines of 1.6, 1.8 and 2.0 litres are offered. The more sporting three-door model with wider track, wheel arch blisters, and a choice of 2.0-litre or 186 bhp 2.0 turbo engines makes a far better case for itself.

Best All-Rounder: Delta 2.0

BODY STYLES: Hatchback
ENGINE CAPACITY: 1.6, 1.8, 2.0, 2.0Turbo, 1.9TD

PRICE FROM: approx. £12,000
MANUFACTURED IN: Italy

Lancia Dedra

There was a time when enthusiasts bought cars like Lancias because of a heightened sense of charm, talent and driver appeal that these Italian marques possessed. Sadly, the Lancia Dedra has none of these fine virtues in sufficient quality to make it appeal to this small market, and this was almost certainly the main reason why Lancia pulled out of the British market in 1995.

Even so, the Dedra lives on in other parts of Europe. Though attempting to compete with the likes of the BMW 3 Series, what it provides is a very ordinary package for the family saloon and estate buyer. Best aspect is the range of 1.6, 1.8 and 2.0-litre engines; all have a certain verve and character, but the turbo version provides fine pace and flexibility. But for the driver that's about it. The driving position is far from acceptable, the gearchange is sloppy, the handling lacks poise and the ride is choppy. When there are so many rivals offering greater talent, the Dedra ranks as one of those cars that never quite made it.

Best All-Rounder: Dedra 2.0

BODY STYLES: Saloon
ENGINE CAPACITY: 1.6, 1.8, 2.0i

PRICE FROM: approx. £14,000
MANUFACTURED IN: Italy

Lancia Kappa

The Kappa is a bigger, sleeker and altogether more convincing flagship for Lancia than the Thema it replaced. Offered with a wide range of engines, including a 200 bhp turbo, a 205 bhp version of Alfa's three-litre V6 and 2.0 and 2.4-litre versions of Lancia's all-new five-cylinder Superfire engine, the Kappa has, on paper at least, what it takes to compete with BMW and Mercedes-Benz. Though still a relatively new design, an S2 Kappa is due for 1997, while the saloon and estate were joined by a coupe in '96.

The new car is markedly more spacious than the Thema and has a neatly-styled interior. In tune with the latest developments from the Fiat group, this big Lancia's chassis dynamics are a sound improvement over past efforts, with the ride and handling now far closer to Europe's best. Lancia has pulled out of Britain, so the IDEA-designed Kappa will never be sold here. But the next big Alfa saloon borrows heavily from the Kappa's styling.

Best All-Rounder: Kappa V6

BODY STYLES: Saloon, Estate, Coupe
ENGINE CAPACITY: 2.0, 2.0 Turbo, 2.4, 3.0V6, 2.4TD
PRICE FROM: approx. £19,000
MANUFACTURED IN: Italy

Land Rover Defender

The old war horse goes on forever. Still popular with farmers and armed forces the world over, the Defender is as tough and rugged as always. Available in two basic forms, with either a 90 inch or 110 inch wheelbase, dozens of derivatives have been developed from this basic steel girder chassis/aluminium panelled platform. So pick ups, cranes and fire engines are built alongside the popular hard top and estate versions.

Before the introduction of the Discovery, Land Rover upgraded the Defender with a 'County' pack in a not altogether successful attempt to civilise it for road use. The trim and equipment may have been improved, but underneath it is still the same, and that means a harsh ride, uncompromising driving position and massive fuel consumption from the V8. But with the turbo-diesel, and the now standard power steering, the Defender remains a highly sensible vehicle for off-road use.

Best All-Rounder: Defender TDi

BODY STYLES: Estate, Convertible
ENGINE CAPACITY: 2.5TD
PRICE FROM: £18,900
MANUFACTURED IN: UK

Land Rover Discovery

The Discovery has stood the test of time remarkably well, remaining the best-selling Land Rover seven years after its introduction. Closely-related to the recently superseded Range Rover, the Discovery brought fashion to the four-wheel-drive market, an area where most other manufacturers are still struggling to catch up. 1997 sees a minor facelift, but the essentials - tall body with roof windows, Terence Conran-designed interior, the option of seven seats and a whole host of life-style accessories - remain. The Discovery's toughest competitor might just turn out to be the new small Land Rover, due late in 1997.

The off-road performance is unsurpassed, although more importantly (95% of buyers never go off-road) the suspension is comfortable on Tarmac roads too Most people seem to like the high, commanding driving position. Of the three engines the 2.0-litre is under powered for the task, the 3.9-litre V8 gives the most satisfying performance, but the favourite - for its economy - is the 2.5 turbo diesel.

Best All-Rounder: Discovery 2.5 TDi

BODY STYLES: Estate
ENGINE CAPACITY: 2.0, 2.5TD, 4.0V8
PRICE FROM: £19,150
MANUFACTURED IN: UK

Land Rover Range Rover

Quite unlike any other 4x4 off-roader, the Range Rover manages to compete equally well with a whole range of luxury products, be they cars, yachts or second homes. But that is not to decry the awe-inspiring off-road ability of the Range Rover, increased still further in the current model with the ability of the suspension to increase the ground clearance. Three engines are on offer, a 4.6-litre V8 for the top-of-the-range HSE, a 4.0-litre V8 and BMW's impressive 2.5-litre, six-cylinder turbo diesel.

A commanding driving position and excellent cabin design help to explain why the Range Rover's appeal extends beyond that of the traditional four-wheel-drive buyer. The interior is luxurious and of a very high quality, with plenty of room for passengers and luggage. But none of this comes cheap, with the V8 and turbo-diesel models starting at £35,000 and that's without a leather interior, air conditioning or automatic gearbox.

Best All-Rounder: 4.0 V8

BODY STYLES: Estate
ENGINE CAPACITY: 2.5TD, 4.0V8, 4.6V8
PRICE FROM: £34,500
MANUFACTURED IN: UK

Lexus GS300

Slightly smaller and a good deal less expensive than the LS400, this four-door saloon with a coupe-like profile, powered by a six-cylinder 3.0-litre engine, ranks alongside the likes of the E-class Mercedes in terms of price. As you might expect from a Lexus, it carries a comprehensive specification and it's meticulously assembled with a reassuringly solid feel.

The sleek profile is in part responsible for the shortage of rear passenger space. This is a car aimed clearly at the driver rather than the chauffeured passenger, and that's underlined by a standard of road behaviour any keen driver will warm to. Safe and wieldy through the turns, the Lexus encourages enthusiastic driving, although at high speed it can't muster quite the same level of iron body control as the best Europeans. If the GS300 is mildly disappointing for its lack of urge, this is made up for by a smoothness both in the engine and the auto transmission.

Best All-Rounder: GS300

BODY STYLES: Saloon
ENGINE CAPACITY: 3.0
PRICE FROM: £33,400
MANUFACTURED IN: Japan

Lexus GS400

The Lexus proved beyond all doubt that a Japanese car manufacturer could produce a luxury car to rival the world's finest. Though at first the Toyota-built LS 400 was looked upon as the poor man's Mercedes-Benz, it soon became clear the Lexus was well able to trade insults with Europe's best.

Those who have driven the Lexus are left with the memory of silky refinement and effortless urge. A super-smooth transmission, an extraordinary ride and thoroughly capable handling mark this out as one of the best of its class. It comes loaded with every imaginable creature comfort – in contrast to some rivals – and the 4.0-litre V8 engine is powerful enough to show 150mph pace. What it lacks is understandable - the sense of heritage that comes as a matter of course with a Mercedes-Benz, Jaguar or BMW. It's all rather clinical, but even so, you would be hard pushed to find a competitor more than marginally better in any respect.

Best All-Rounder: LS 400

BODY STYLES: Saloon
ENGINE CAPACITY: 4.0V8
PRICE FROM: £46,000
MANUFACTURED IN: Japan

Lincoln Continental

If this luxury-liner Lincoln hadn't already got enough gizmos, '97 sees it equipped with a Remote Emergency Satellite Cellular Unit, or RESCU, package. Using voice-activated cellular phone technology and two console-mounted buttons, the driver can summon roadside assistance or emergency help. A Global Position System directs rescue services to where the Lincoln is parked, or points you in the right direction when you're lost. The package also includes Michelin run-flat tyres.

V8 muscle, in the form of Ford's formidable 4.6-litre, 260 bhp V8, provides the Continental with muscular performance and whisper-quiet cruising. Electronic wizardry gives the driver the choice of firm, normal or plus suspension settings, along with three choices of steering weight. Both work best in the middle settings. Despite its bulk, the Continental is an accomplished handler and provides a supple, absorbent ride for those inside its luxurious leather and wood-lined cabin.

Best All-Rounder: Continental

BODY STYLES: Saloon
ENGINE CAPACITY: 4.6V8
PRICE FROM: $42,000
MANUFACTURED IN: US

Lincoln Mark VIII

Lincoln's Ab-Fab luxury coupe gets a bold new look for '97. Striking high-intensity discharge headlights - seemingly twice the size of the last ones - flank a new grille that appears to start half way up the bonnet and droops way down into the front bumper. If Ford wants people to notice the new Mark VIII, this is the way to do it. There's a similar, aggressive treatment at the rear, with bold new lights that includes a neon strip that creates a 3D glow.

Mechanically, the new 'Mark' sticks with its 280 bhp 4.6-litre V8 and quick-changing four-speed auto. This sleek, well-proportioned coupe is certainly a fast mover for its size, scorching to 60 mph from standstill in around seven seconds. For those who want even greater athleticism, the LSC version packs 10 more horses and comes with beefier anti-roll bars and firmer damping. Inside, the '97 Mark VIII gets a designed facia and re-shaped seats.

Best All-Rounder: Mark VIII Coupe

BODY STYLES: Coupe
ENGINE CAPACITY: 4.6V8
PRICE FROM: $40,000
MANUFACTURED IN: US

Lister Storm

The number of front-engined supercars can be counted on one hand, but Lister has joined the few with its spectacular Storm. It uses a development of Jaguar's V12 tucked low down behind the front wheels, with the capacity increased to 7.0 litres and the addition of twin superchargers. The result is almost 600 bhp and over 200 mph in sixth gear. But it's not just the performance that's impressive. The Storm is a dazzling combination of aluminium honeycomb chassis, carbon fibre composite panels, with steel and aluminium used in the doors for strength and safety.

The benefits of keeping the engine at the front of the car, rather than behind the seats as with most supercars, is the increased practicality. The Storm offers seating for four and has room for golf clubs and trolley in the boot. Not surprisingly it is an expensive proposition, but an abundance of Connolly leather and Wilton carpeting help establish this as a true high performance luxury supercar.

Best All-Rounder: Storm 7.0 V12

BODY STYLES: Coupe
ENGINE CAPACITY: 7.0V12
PRICE FROM: £220,000
MANUFACTURED IN: UK

Lotus Elise

The Elise brings Lotus back to its roots, a minimalist sports car which relies more on its light weight than a high powered engine for its performance. Technically it is a fascinating car, the enormously strong epoxy-bonded aluminium chassis weighs just 70 kg, aluminium composite brakes are used for the first time in a production car, the driver's seat is mounted closer than the passenger's to the centre line of the car to improve weight distribution. Power is provided by Rover's MGF engine mounted behind the seats.

The Elise is a basic car in the extreme, with rudimentary hood, no carpets and lots of bare aluminium. But its primary, perhaps only, purpose is to provide an unrivalled driving experience and that it does with utter conviction. The engine may not be that powerful but the lack of weight ensures the Elise is very, very quick. Better still are the steering and handling where the Elise is almost unsurpassed.

Best All-Rounder: Elise 1.8

BODY STYLES: Convertible
ENGINE CAPACITY: 1.8

PRICE FROM: £20,000
MANUFACTURED IN: UK

Lotus Esprit

Britain's most sporting of sports car manufacturers has been going through some difficult times. Part of the problem has been a reliance on a single model, the Esprit, which has had an increasingly tough time competing with likes of Ferrari and Porsche, both of which employ far more appealing engines than Lotus's four-cylinder turbo unit. Now things are looking up for Lotus, its product line doubled with the exciting new Elise, and an all-new V8 engine for the Esprit.

With its twin turbochargers the 3.5-litre V8 produces an impressive 350 bhp, enough to move the Esprit's performance into a totally new league. The rest of the car is much as before, with a steel backbone chassis clothed in a stylish composite plastic body. The steering, handling and ride remain a Lotus strength, but gearchange is a weak point. Lotus has also introduced the GT3, an entry-level Esprit based around the racer, with 2.0-litre turbo engine.

Best All-Rounder: Esprit V8

BODY STYLES: Coupe
ENGINE CAPACITY: 2.2Turbo, 3.5V8

PRICE FROM: £49,000
MANUFACTURED IN: UK

Marcos Mantara

The Marcos is a 1960s sports car that has evolved continuously over three decades – and it has thrived. Originally styled by Dennis Adams with a fibreglass body, a four-cylinder engine and wooden aircraft-type structure, the current evolution now has a steel spaceframe chassis, a meaner body and a meaty V8. There's also an open-top version. But the overall philosophy remains; a high-performance small-production sports car with style and character.

Much of the original Marcos style is carried over to the Mantara, but the LM is based on the Marcos racers and has the full steroid treatment. Both have either a 4.0-litre 190bhp or 5.0-litre 320bhp V8; one is merely quick, the other is a rocket in a car wieghing little more than a Fiesta; it demands respect. The chassis is up to it though and the classic leather and Wilton-trimmed cabin with hip-hugging seats and shoulder-high central tunnel make the owner feel as if he's driving something very special.

Best All-Rounder: Marcos LM400

BODY STYLES: Coupe, Convertible
ENGINE CAPACITY: 4.0V8, 5.0V8

PRICE FROM: £28,000
MANUFACTURED IN: UK

Maserati Ghibli

Maserati's range has been reduced to two models, of which the Ghibli is the one which has the task of continuing the high-performance GT image of the past. It answers the performance question with conviction, a choice of three twin-turbo V6 engines providing explosive bursts of power once the turbochargers start doing their job. The least powerful is the 2.8 which produces 284 bhp, with the standard 2.0 managing 306 bhp; both cars accelerate to 60 mph in around 5.5 seconds and will reach 165 mph.

There's also a special edition Ghibli Cup, based around the race car, which has an astonishing 330 bhp. The advantage of the Cup is that carbon fibre is used in place of the overwhelming wood which dominates the interior of the standard cars. The Ghibli has been improved for '97, with a new multi-link rear suspension and a six-speed gearbox. For all its faults and foibles, the Ghibli is a car with which you could so easily form a lasting relationship.

Best All-Rounder: Ghibli 2.8 V6

BODY STYLES: Coupe
ENGINE CAPACITY: 2.0V6 Turbo, 2.8V6 Turbo
PRICE FROM: £45,000
MANUFACTURED IN: Italy

Maserati Quattroporte

It's a strange beast, the Maserati Quattroporte. Like so many high-performance Italian cars, it's an infuriating mix of things you passionately love and things you couldn't possibly live with. Designed as the luxury express that tops the Maserati range, the four-door Quattroporte certainly has all the trimmings. Yet it is let down by incomprehensible switchgear, a flawed driving position and a cramped rear compartment. And, crucially, cars costing half the price are better built.

Yet the Quattroporte makes up for it in part with glorious engines, with the 2.0 and 2.8-litre V6s recently joined by a 3.2 V8. All have their power boosted by twin turbochargers and all have similar characteristics - unremarkable acceleration at low revs, then an explosion of performance once the turbos start working. The results are highly impressive with the six-speed manual gearchange, less so with the four-speed automatic transmission that most buyers are said to favour.

Best All-Rounder: Quattroporte V8

BODY STYLES: Saloon
ENGINE CAPACITY: 2.0V6 Turbo, 2.8V6 Turbo, 3.2V8 Turbo
PRICE FROM: £57,700
MANUFACTURED IN: Italy

Mazda 323

This Escort-sized hatchback is certainly distinctive, especially in five-door trim. But only when Mazda's 323 is displayed in top-spec 2.0 V6 form with its larger wheels does the designer's intent shine through. In every other form, the spindly wheels are lost within those high-sided flanks. The ugly three-door hatch and more conservatively styled saloon have been dropped from the UK-line up, leaving just the five-door with four-cylinder engines of 1.5 and 1.8-litres, plus the fast and refined 2.0 V6.

Whatever your views on the styling, few will argue about the 323's generally capable and likeable character. There is a typically Japanese slickness to the controls and a well planned and comfortable driver environment. It grips well handles capably and has good-natured engines. A high specification and excellent safety credentials add to the appeal, but those travelling in the rear might have something to say about the lack of headroom, and the poor visibility.

BODY STYLES: Hatchback, Saloon
ENGINE CAPACITY: 1.3, 1.5, 1.8, 2.0V6
PRICE FROM: £12,000
MANUFACTURED IN: Japan

Mazda 626

Mazda's interpretation of the mainstream family car is not a bad effort . It comes either as a five-door hatch or as a saloon, with 1.8-litre or 2.0-litre; a 2.5-V6 was recently dropped in the UK. The 626 was one of the first of its type to feature the smoothly rounded styling that's now so familiar, and it's a particularly roomy family car, well put together and nicely finished. Rear passenger space is particularly good and the light controls and power steering make it an easy car to drive.

Although the 626 is relatively softly sprung with body roll through the turns, it rides well at speed and the smooth and willing engines endow it with more driver appeal than might be expected. If anything lets it down, it's the dull interior, with large expanses of hard uninviting plastic, but in most respects the 626 forms a very positive overall impression. Mazda will be introducing an updated model sometime in 1997.

Best All-Rounder: 626 1.8 LXi

BODY STYLES: Hatchback, Saloon
ENGINE CAPACITY: 1.8, 2.0, 2.5V6, 2.0D
PRICE FROM: £12,600
MANUFACTURED IN: Japan

Mazda Xedos 6

The Xedos 6 is the car that disproves the oft-held belief that the Japanese don't build cars with true character. With its distinctive yet elegant shape and its silky six-cylinder engine, the Xedos 6 is a star player in the compact executive class. It gives little away to the likes of BMW and Audi for showroom appeal and even has their measure in many respects.
A 1.6-litre four-cylinder version is available but it's the 2.0 V-6 that lies at the core of the Xedos 6's appeal.

The V6 provides storming performance when you want it yet with creamy refinement at all times. An excellent gearchange together with capable handling and ride add to the appeal. But it's the stylish and well-built interior that makes this car feel just that little bit special. Passengers don't get such a good deal, however. The Xedos 6 is cramped in the back and the boot space is small.

Best All-Rounder: 2.0i Sport

BODY STYLES: Saloon
ENGINE CAPACITY: 1.6, 2.0V6
PRICE FROM: £16,000
MANUFACTURED IN: Japan

Mazda Xedos 9

Mazda has proved with the Xedos 9 it has the ability to convincingly break the mould set by Europe's grand masters. Armed with a stout 168bhp V6 engine that offers tremendously refined progress and with overall levels of silence that would do justice to a Jaguar, the impressively roomy and generously equipped Xedos 9 does all the right things on a purely practical level. It's tightly and accurately assembled too, and it carries the assurance of plenty of safety equipment.

What it fails to do is to provide the sense of occasion that you would get with a Mercedes or a BMW. Nor does the engine and super-smooth auto transmission offer quite the level of pace expected of a 2.5-litre engine. So although the Xedos 9 comes comfortably close to Germany's finest, and at a price that looks highly competitive given the extensive specification, Mazda's executive contender misses the mark by a fraction.

Best All-Rounder: Xedos 9 SE

BODY STYLES: Saloon
ENGINE CAPACITY: 2.5V6
PRICE FROM: £27,000
MANUFACTURED IN: Japan

Mazda MX-3

The MX3 certainly sets itself apart from the common herd, crossing the line between cute coupe and alternative hot hatchback. It defies convention not just in its ovoid appearance, but also in boasting one of the smallest and sweetest V6 engines yet conceived. If, however, you go for the alternative power unit, a 1.6-litre four-cylinder engine, you are compelled to take automatic transmission. The combination is far from successful, with sluggish performance which belies the MX-3's sporting appearance.

While it is possible to accommodate four people within the MX-3, the restricted space means the driver should expect the occasional whinge from those in the back. The interior is disappointingly conventional in that Japanese, plasticky way, but consolation comes in the form of a large-sized helping of driver appeal. The chassis delivers crisp responses although anyone expecting serious performance from the V6 engine will be disappointed.

Best All-rounder: MX-3 V6

BODY STYLES: Coupe
ENGINE CAPACITY: 1.6, 1.8V6
PRICE FROM: £14,800
MANUFACTURED IN: Japan

Mazda MX-5

Mazda can justifiably take the credit for kick-starting the sports car market after it went into the doldrums following the demise of the MGB and Midget. Since its introduction in 1988, the MX-5 has been a phenomenal success world-wide. In maintaining a no-frills approach to the traditional front-engine, rear-drive concept, Mazda managed to create a modern classic that's affordable and fun.

That said, anyone expecting scintillating thrust is likely to be disappointed. The 1.6 develops a meagre 88 bhp though the 1.8 has a more encouraging 130 bhp. Both produce an engaging yowl at the top end, but many a hot hatch has the legs to blow them away. So it's a relief that straight-line pace is not what the MX-5 is all about. It's about pure driving pleasure. The crisp steering responses, the exquisitely balanced cornering, the flick-switch gearshift and the reassuring feel of the brakes all add up to a car that flatters the driver and produces a warm glow all over.

Best All-Rounder: MX-5 1.8i

BODY STYLES: Convertible
ENGINE CAPACITY: 1.6, 1.8
PRICE FROM: £13,500
MANUFACTURED IN: Japan

Mazda MX-6

With the likes of Fiat and Alfa bringing new impetus with their new sports coupes, this class is full of vitality right now. One thoroughly worthy, well made and characterful coupe that has remained at the core of this category for some years now is Mazda's MX-6. It's a well made, refined and fast contender with a silky V6 engine and good practicality too. Folding rear seats extend the boot space and it seats four - though only grudgingly.

The MX-6's character is very much in the grand tourer mould, which is to say there are faster, better handling coupes around. But the MX-5 combines a strong range of virtues to make it both comfortable and refined as well as having a good measure of pace and handling prowess. It's well equipped, and the fit and finish of the panel work is first rate and it is comfortable and easy to drive. The real downside, as with so many Mazdas, is that the interior lacks class, with too much use of uninspiring plastic.

Best All-Rounder: MX-6 2.5i GT

BODY STYLES: Coupe
ENGINE CAPACITY: 2.5V6
PRICE FROM: £20,300
MANUFACTURED IN: Japan

McLaren F1

Britain is the centre of the motor racing industry so it was only a matter of time before that huge background of expertise was utilised to produce a road car which was simply the most advanced available anywhere. The McLaren F1 is that car, a truly unique vehicle which incorporates several world firsts for a road car: a fully carbon fibre composite monocoque structure; fully active fan-assisted ground effect aerodynamics; central driving position with two offset rear passenger seats. It also sets a record for the most expensive off-the-shelf-car - £634,500!

Designer Gordon Murray incorporated many years of motor sport experience into the F1, making use of frighteningly expensive materials in order to keep the weight down. The engine comes from BMW, and performance is unbelievable - acceleration to 60 mph in 3.2 seconds, 230 mph maximum speed. It was good enough to win at Le Mans in '95.

Best All-Rounder: McLaren F1

BODY STYLES:	Coupe	PRICE FROM:	£635,000
ENGINE CAPACITY:	6.1V12	MANUFACTURED IN:	UK

Mercedes-Benz C-Class

The C-Class may be Mercedes-Benz's smallest and cheapest saloon, but the fact that it retains the same high levels of engineering and quality as any other car produced by the German manufacturer ensures it is an attractive proposition. 'Cheap', however, is a relative term, for £20,000 is the starting point, and that's before the obligatory stereo and automatic transmission. 1996 saw the first significant changes to the C-Class, with the introduction of an estate car, new 16-valve engines and a turbo-diesel. And the Kompressor is the first supercharged Mercedes for more than 50 years.

The C-Class excels in areas other than its quality - ride, noise levels and refinement particularly. Most drivers will be comfortable in the firm front seats, though they have to put up with a large steering wheel and, on many models, a minimum of features. Room in the back isn't great either, but if this isn't a worry, the C-Class will provide year's of sterling service.

Best All-Rounder: C200 Elegance

BODY STYLES:	Saloon, Estate	PRICE FROM:	£19,250
ENGINE CAPACITY:	1.8, 2.0, 2.3K, 2.8, 3.6, 2.2TD, 2.5TD	MANUFACTURED IN:	Germany

Mercedes-Benz E-Class

Mercedes-Benz has become more adventurous with its styling, the latest E-Class executive saloon now unmistakable with its four round headlights. Beneath the surface, though, its the same strong attributes we've come to expect, legendary build quality which ensures the cars will knock off 100,000 miles in their stride and a high level of safety, which includes forward and side airbags for those in the front seats. The saloon was been joined by an estate for 1996, though initially only with the smaller 2.0 and 2.3-litre engines; saloons get the option of 2.8 and 3.2 six-cylinder engines as well as two diesels.

Unlike the BMW 5 Series, its main competitor, a six-cylinder engine comes a long, long way up the price ladder on the E-Class. The 2.0-litre car is rather lifeless, but the 2.3 four-cylinder is refined with good levels of performance. This E-Class is more enjoyable to drive than ever, although the downside is that the ride is noticeably firmer.

Best All-Rounder: E230

BODY STYLES:	Saloon, Estate	PRICE FROM:	£24,000
ENGINE CAPACITY:	2.0, 2.3, 2.8, 3.2, 4.2V8, 5.0V8, 2.2TD, 2.9TD	MANUFACTURED IN:	Germany

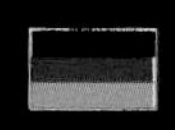

Mercedes-Benz S-Class

Mercedes—Benz was out to prove a point when it developed the S-Class; to make the best luxury car in the world, bar none. In so doing the development costs went way over budget, it acquired a hefty paunch, and when the S-Class eventually appeared in a rather more environmentally-aware society than when conceived, it was derided for its conspicuously extravagant waste of the world's resources.

Two body styles are offered - a roomy saloon or a positively cavernous long-wheelbase version. In six-cylinder guise it will barely pull its two tons along, but in V8 or V12 form it will storm along like a sports car while devouring fuel almost as fast as you could pour it in. And yet none of this can detract from this vast car's towering achievements. For absolute automotive refinement and comfort, the likes of BMW, Jaguar and Lexus must give way. Engineering excellence, if not elegance, comes no better.

Best All-Rounder: S320

BODY STYLES:	Saloon, Coupe	PRICE FROM:	£41,000
ENGINE CAPACITY:	2.8, 3.2, 4.2V8, 5.0V8, 6.0V12	MANUFACTURED IN:	Germany

Mercedes-Benz SL

The SL sports car still looks as fresh and enticing as the day it was launched back in 1989. Being a Mercedes-Benz, part of the attraction is the rock-solid build quality which means buyers have confidence the car will just go on and on. Then there's undeniable grandeur of driving around in Mercedes-Benz's most prestigious vehicle. The car contains a host of high-tech features, including a roll-over bar which pops up in a fraction of a second should the car invert, as well as, naturally, a fully automatic hood. Engine choice ranges from a 2.8-litre six to two six-litre options, a V8 or V12.

To drive, the SL is as good as it looks. Though it may find much favour with wealthy lady drivers, the SL is a truly sporting car, with a very well-developed chassis and strong performance. The V12 is over the top, both in price and weight - the 500SL V8 is as quick and much cheaper. Realistically, the six-cylinder models give the SL experience with little loss in everyday performance.

Best All-Rounder 320 SL

BODY STYLES:	Convertible	PRICE FROM:	approx. £57,000
ENGINE CAPACITY:	2.8, 3.2, 5.0V8, 6.0V12	MANUFACTURED IN:	Germany

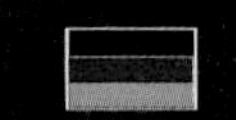

Mercedes-Benz SLK

The new 'poor-man's' Mercedes SL hit the streets towards the end of 1996 with an order book already well into 1998. Similar in size to BMW's Z3, but built in Germany which immediately gives it a price disadvantage, buyers are still flocking to buy this exciting new sports car. With prices starting at around £25,000, the SLK offers desirability by the bucket load at half the cost of an SL.

Key features are just two seats, front-engine rear-wheel-drive and the most innovative roof system yet on a sports car. The SLK has a steel roof which divides and lowers electrically into the boot. Roof up, there are all the benefits of a real coupe, down, it's a sports car. Except that with a traditional large steering wheel and poor seat bucketing, the emphasis is more towards comfortable cruising than outright sports-car motoring. There's also the problem that the stored roof takes 60% of the boot space. Power comes from a either a 2-litre or 2.3 supercharged engine.

Best All-Rounder: SLK 2.3

BODY STYLES:	Convertible	PRICE FROM:	approx. £23,000
ENGINE CAPACITY:	2.0, 2.3 Supercharged	MANUFACTURED IN:	Germany

Mercedes-Benz V-Class

In European showrooms from autumn '96, the V-class brings Mercedes-Benz into the blossoming MPV market. A 'play room, living room and saloon car rolled into one' is the claim, but the ingredients are familiar MPV with some special Mercedes' touches. Each of the individual seven seats has its own lap-and-diagonal safety belt and headrest, making this potentially the safest MPV. Unlike most other competitors, the seats fold away within the V-Class, meaning you don't have to leave the seats at home when you need the extra luggage capacity.

This is the first Mercedes-Benz to be built outside Germany - in Vitoria Spain. Two engines are available initially - 2.3 petrol and 2.3 turbo-diesel; a 2.8 V6 Volkswagen-based engine will follow later. Instead of steel springs the rear suspension uses air bellows which are inflated according to how heavily loaded the car is, giving better comfort and car control.

Best All-Rounder: Too soon to say

BODY STYLES: Multi Purpose Vehicle
ENGINE CAPACITY: 2.3, 2.8V6, 2.3TD
PRICE FROM: £22,000
MANUFACTURED IN: Spain

Mercury Mountaineer

With a new chrome grille, different lights at the rear end and some new badge work, Ford's mega-successful Explorer is morphed into a Mercury Mountaineer for the Blue Oval's upscale Lincoln-Mercury division. Despite the minimal changes, it's nevertheless an appealing package; standard 4.9-litre V8 coupled to an automatic four-wheel drive system that kicks-in only when needed, and a nicely kitted-out interior with lots of leather.

As over 90 per cent of Mountaineer owners will never venture off-road in this Explorer-clone, civilised on-road manners were vital. Hence the Mercury rides smoothly, steers accurately and doesn't roll on corners like a tramp steamer in a Force Five. The big V8 offers more torque than performance and does a swell job of punching the Mountaineer up steep slopes and out of tight bends. It's gets a little breathy when revved but stays refined and muted at cruising speeds.

Best All-Rounder: Mountaineer 4x4

BODY STYLES: Estate
ENGINE CAPACITY: 5.0V8
PRICE FROM: $30,000
MANUFACTURED IN: US

MG MG*F*

It's been a long time coming, this first all-new MG sports car for over 30 years, and it has hit the streets at the same time as a clutch of other new roadsters, but the MGF can hold its head high. Not least because it is the least compromised of all these sports cars in terms of its mechanical specification. The MGF is mid-engined which theoretically provides the best possible handling, and it's packed with technical niceties like electric power steering and variable valve timing on the fastest VVC model.

It all works brilliantly well, too. The performance from either of the 1.8-litre engines is sufficient to provide a big grin, but the real surprise is the quality of the suspension, which ensures stunning road holding yet at the same time provides a comfortable ride. The interior is well-trimmed, the hood easy to use and there's a fair-sized boot behind the engine compartment. All-in-all a great fun package.

Best All-Rounder: MGF 1.8

BODY STYLES: Convertible
ENGINE CAPACITY: 1.8, 1.8VVC
PRICE FROM: £16,400
MANUFACTURED IN: UK

Mitsubishi Colt

Mitsubishi's fifth generation Colt went on sale early in 1996. With a price straddling the Fiesta and Escort, the new Colt is 75mm shorter than before, yet keeps the same passenger and luggage space. Available only as a three-door hatchback, there's a choice of 1.3 or 1.6-litre petrol engines with the option of a sophisticated automatic gearbox. The more recently introduced Mirage offers the trimmings of a GTi - alloy wheels and so on - with the standard 1.6 engine.

There's lots to like about the Colt. It is supremely easy to drive, with power steering, slick gearchange and good visibility. Performance even from the 1.3 is eager and, for those in the front at least, comfort is above average. The bugbear is room in the back - worse than many of the better superminis. If that's not a problem, the Colt makes an appealing package.

Best All-Rounder: Colt 1.3 GLX

BODY STYLES:	Hatchback	PRICE FROM:	£9,800
ENGINE CAPACITY:	1.3, 1.6, 1.8	MANUFACTURED IN:	Japan

Mitsubishi Galant

Mitsubishi has had to think again about its Galant. Its position as the mainstream family car has been taken by the European-built Carisma, so the new 1997 model has received a kick upmarket. Thus the Galant is now a touch larger and is being pitched against executive rather than family cars. Sales began in Japan in September 1996, but the Galant does not arrive in Europe until the spring of '97.

Two body styles are offered, four-door saloon and estate, with front wheel drive and the option of manual or 'intelligent' automatic transmission. European buyers get the choice of two petrol engines, a 2.0-litre 135 bhp four-cylinder and a 2.5-V6 with 160 bhp as well as a 2.0 turbo-diesel. Japanese buyers also get the option of a revolutionary 1.8-litre direct injection petrol engine. The new technology incorporated in this design is claimed to combine the power output of a petrol engine with the economy of a diesel.

Best All-Rounder: Too soon to say

BODY STYLES:	Saloon, Estate	PRICE FROM:	£18,000
ENGINE CAPACITY:	1.8, 2.0, 2.5V6, 2.0TD	MANUFACTURED IN:	Japan

Mitsubishi Carisma

Mitsubishi's first European manufactured car is built in the Netherlands alongside the Volvo S40, with which it shares a number of basic design points. Despite this European emphasis, this Mondeo-class contender was actually designed in Japan. Engines are 1.6 and 1.8 petrol and a diesel; the five-door hatchback has recently been joined by a saloon.

With an air of quality and high equipment levels, the Carisma is something of a bargain in its first year. Well-bucketed seats keep those in the front comfortable although the material on the cheaper versions is decidedly tacky; space in the rear is average, no better. The 1.6 is smooth but needs working hard to get reasonable performance, making the Carisma disastrously slow when coupled to the auto gearbox. The 1.8 is a much better bet, although the problem of an insufficiently smooth ride and noise from a number of sources place it firmly in the midfield of family cars.

Best All-Rounder: Carisma 1.8 GLX

BODY STYLES:	Hatchback, Saloon	PRICE FROM:	£11,000
ENGINE CAPACITY:	1.6, 1.8, 1.8DOHC	MANUFACTURED IN:	Netherlands

Mitsubishi 3000 GT

A Spyder convertible model joined the Mitsubishi 3000 GT range for 1996, while the latest mechanical addition to this technological showcase of a car is a six-speed gearbox. With the existing four-wheel drive, twin-turbo V6 24-valve engine, and all the bells and whistles that anyone could wish for, the 3000 GT is surely the ultimate expression of sports car complexity. But you can't help wondering whether it's all just a bit *too* complicated, with a TV screen for the heater settings and an audio system that takes a science degree to operate.

There's no denying the 3000GT, with its 281bhp engine and all-weather grip, makes a hugely capable and brutally swift Tarmac scorcher. The all-wheel drive adds a sure-footed confidence to its progress yet there's a feeling that the whole car is too heavy, too soft and too complex to fulfil its role as a truly responsive and tactile driving machine. Think of it as a trans-continent express rather than as a 911-eater.

Best All-Rounder: 3000 GT

BODY STYLES: Coupe, Convertible
ENGINE CAPACITY: 3.0V6 Turbo
PRICE FROM: £44,000
MANUFACTURED IN: Japan

Mitsubishi Space Wagon & Runner

Smaller than most MPVs, the Space Wagon is a purpose-built seven-seater powered by a smooth 2.0-litre 16v 'bal-ancer-shaft' engine. The compact size and powerful engine means it has the advantage of being much easier to drive and park around town, as well as providing more eager performance than competitors. The front seats are very comfortable, in the middle row there is yards of legroom, but things are cramped in the rear and only really suitable for children. There's little boot space with all three rows in use, but rows two and three can be quickly folded away to give a large luggage area.

Renault might like to think it has invented a new type of vehicle with the Megane Scenic, but Mitsubishi has been offering its Space Runner since 1991. Similar to the Wagon from the front seats forward, the trim is sportier and the engine a lively 1.8-litre. The overall length is 225 mm less, which means only one row of rear seats. An unusual feature is the single sliding rear passenger door.

Best All-Rounder: Space Wagon 2.0

BODY STYLES: Multi Purpose Vehicle
ENGINE CAPACITY: 1.8, 2.0, 2.0TD
PRICE FROM: £14,300
MANUFACTURED IN: Japan

Mitsubishi Pajero Junior

As successive generations of off-roaders get longer, heavier and more powerful, so new opportunities open up for smaller vehicles. Suzuki and Toyota have shown that there's a mar-ket for more compact recreational vehicles, and now Mitsubishi has gone two-steps further than anyone with its tiny Pajero Junior. Designed to look like a scaled-down Shogun, the Junior gets a colour-keyed grille, big front bumper, flared wheel arches and chunky tyres to give it the full-off-road look. It has genuine off-road capability too, with selectable four-wheel-drive coupled to a high and low ratio gear ranges.

Power is provided a 1.1 litre 16-valve engine which, despite its impres-sive 79 bhp, is not going to pull much of a load through mud. More the Junior is aimed at looking good rela-tively cheaply, and it has the equip-ment levels to prove it - air condition-ing, power steering, drive's air bag and ABS brakes - are all standard.

Best All-Rounder: Pajero ZR-II

BODY STYLES: Estate
ENGINE CAPACITY: 0.7, 1.1
PRICE FROM: n/a
MANUFACTURED IN: Japan

Mitsubishi Shogun

Immensely successful world-wide, the Shogun has acquired a fine reputation as the car that keeps on going where others give up. For many it is the archetypal Japanese off-roader, with an extensive range covering short and long wheelbase body styles. All Shoguns are distinguished by Mitsubishi's sophisticated selectable four-wheel drive system with engines chosen from two excellent turbo-diesels and two powerful if thirsty petrol V6s.

The Shogun's interior may suffer from semi-shiny plastic and tacky switchgear but there's no denying its durability. That toughness extends to the running gear too; a solid-looking chassis and beam rear axle underpin the hefty body structure. It adds up to a car that is at home off road as it is on Tarmac. A surprisingly absorbent ride helps to make it a car that's easy to live with, but only the lighter three-door manages to hide its weight and bulk on the road. Up to seven can be accommodated in the five-door.

Best All-Rounder: Shogun 2.8TD

BODY STYLES: Estate
ENGINE CAPACITY: 2.5TD, 2.8TD, 3.0V6, 3.5V6
PRICE FROM: £18,900
MANUFACTURED IN: Japan

Mitsubishi Space Gear

If there was a competition for the biggest MPV, the Mitsubishi Space Gear would be a grand finalist. It is positively huge - so big it forms a basis of a large van as well. The floor is completely flat from front footwell to the tail end, allowing passengers to move around easily. Seven or eight seats are available, with the rearmost flipping up to the sides when not required to free more luggage space.

Unusually for this type of vehicle, the Space Gear has an in-line front engine driving the rear wheels. Four-wheel-drive is an option, using the system from the Shogun which, Mitsubishi claims, gives the Space Gear outstanding all-terrain performance. Engines can be selected from a list of four - 2.4-litre four-cylinder, 3.0 V6 or one of two turbo diesels, 2.5- or 2.8-litre. Top models get electronically controlled suspension. It seems unlikely the Space Gear will be sold in the UK despite its popularity in Europe.

Best All-Rounder: 2.8 TD

BODY STYLES: Multi Purpose Vehicle
ENGINE CAPACITY: 2.0, 2.4, 3.0V6, 2.5TD, 2.8TD
PRICE FROM: approx. £17,000
MANUFACTURED IN: Japan

Mitsubishi FTO

Name apart, the FTO (it stands for Fresh Touring Origination) is a thoroughly exciting coupe with particularly distinctive styling. While the interior is pretty much standard Japanese coupe, with too much black plastic to provide great appeal, the real interest lays in the mechanics of the car, particularly the top of the range model with automatic transmission. Though it is built in right-hand-drive for Japanese markets, it's unlikely we'll ever see the FTO in the UK.

The 2.0-litre V6 has variable valve timing which gives it a Jekyll and Hyde character. Below 6,000 rpm it feels pretty ordinary, but take it through to 8,000 rpm and there's a wail as the 200 bhp is unleashed. A five-speed gearbox is standard but there's also a very clever automatic. This 'learns' your driving style and will change down a gear automatically as the brakes are applied for a corner. In 'sport' mode gears are selected manually by flicking the lever back and forth. Humbler FTOs are available with a 1.8 or lower powered V6.

Best All-Rounder: FTO 2.0 GPX

BODY STYLES: Coupe
ENGINE CAPACITY: 1.8, 2.0, 2.0 Supercharged
PRICE FROM: n/a
MANUFACTURED IN: Japan

Nissan 200 SX

Introduced only in 1995, the 200SX receives its first facelift for the 1997 model year. A revised front end, interior trim and dashboard are the main changes, but the essentials remain the same - that of a rear-wheel-drive 'touring coupe' with a very powerful 2.0-litre turbocharged engine.

And with 200 bhp the performance is exceptional, the maximum speed little short of 150 mph and acceleration to 60 mph in around 7.5 seconds. The quality of the steering and handling make the 200SX a great car to drive although if you want to make the most of the acceleration the engine becomes thrashy and unappealing. There's transmission jerkiness too and the ride is unsophisticated. Inside, the standard 200SX is dull and unenticing, although the leather in the Touring version livens things up a bit. A couple of tiny seats are positioned in the rear, but this Nissan works better when considered as a roomy two-seater

Best All-Rounder: 200 SX

BODY STYLES:	Coupe	PRICE FROM:	£19,700
ENGINE CAPACITY:	2.0, 2.0 Turbo	MANUFACTURED IN:	Japan

Nissan Serena

The Serena may be one of the more compact and least costly of people-carriers, but it's also one of the most successful at packing a lot of people into a small space. Up to eight can be accommodated on three rows of seats within its tall and ungainly shape. It's expecting too much for either the 1.6 or the 2.3 diesel to haul it all along with any vigour, but the 2.0-litre manages successfully to produce a decent pace without effort. It's straightforward to drive too, with light and easy controls.

Facelifted for the summer of 1996, the main improvements are improved levels of safety equipment. Like so many of its ilk, the Serena has precious little space for luggage if all three rows of seats are in use. The Serena is just about roomy enough for the intended task, provided the journey isn't too long and you don't mind rubbing shoulders with your neighbour, but the ride comfort is a touch too nautical for some to bear, especially in the rear.

Best All-Rounder: Serena 2.0 SLX

BODY STYLES:	Multi Purpose Vehicle	PRICE FROM:	£14,500
ENGINE CAPACITY:	1.6, 2.0, 2.3D	MANUFACTURED IN:	Spain

Nissan Patrol

Off roaders don't come much beefier than this. Big and chunky, with a torquey 4.2-litre six-cylinder petrol engine or a 2.8 turbo-diesel, the Patrol looks and feels as though it really means business. The tough separate chassis construction is as expected but a basic four-wheel drive system marks the Patrol as a low-tech contender compared with a Discovery or a Shogun. Inside the dated styling is a major disappointment, while the three-door Patrol is surprisingly cramped in the rear. The long-wheelbase version seats up to seven and the standard equipment tally is a generous one.

On the road, the petrol-engined Patrol encourages a gung-ho enthusiasm, its sheer size intimidating some drivers and most other road users. The storming engine and well controlled cornering endows it with a more car-like driving feel than many of its rivals, but the harsh ride and the dismal economy counts against. Which make the diesel the more sensible, if slower, choice.

Best All-Rounder: Patrol 2.8 TD SE

BODY STYLES:	Estate	PRICE FROM:	£20,500
ENGINE CAPACITY:	2.8TD, 4.2	MANUFACTURED IN:	Japan/ Spain

Nissan Terrano II/ **Ford** Maverick

Trim and minor detailing apart, the Terrano II and Ford Maverick are identical, built in Spain by Nissan with styling by the Italian Michelotti. The idea was to produce an off-road type vehicle biased towards road use to appeal to the estate car or saloon buyer. As such, it's smoother styled and less macho than most of its ilk, with an interior that has a familiar car-like look about it.

It's not an approach that has achieved universal success. The none-too inspired design that offers neither the ride comfort and refinement of an estate nor the macho style of a less self-conscious mud-plugger. Revisions to the styling in mid-'96 have addressed the latter problem to an extent, though the Ford's bold chrome grille will not be to everyone's taste. Two body styles are on offer, a five-door seven-seater and a more cohesive-looking three-door. The 2.7-litre turbo diesel now has 25% more power and is much improved, but the 2.4 petrol engine fails to impress.

Best All-Rounder: Terrano II 2.7TD

BODY STYLES: Off-Roader
ENGINE CAPACITY: 2.4, 2.7TD
PRICE FROM: £16,700
MANUFACTURED IN: Spain

Nissan Skyline

Sold exclusively in Japan and Australia, the Skyline may look like countless other coupes bedecked with wings and spoilers, but beneath all the glitz is a true supercar, albeit in the minor league. Although available as a stylish saloon as well as a cooking two-door, the Skyline which attracts all the attention is the GT-R. Packed full of high-tech features, it's a showcase for Nissan's engineering expertise that has proved its worth both in large capacity touring car racing as well as at Le Mans.

The 2.6-litre six-cylinder engine is equipped with 24 valves and twin ceramic turbochargers, resulting in an output of 280 bhp and a top speed in the region of 160 mph. The power goes through limited slip differentials to all four wheels, and there's four-wheel-steering too, just to help keep everything pointing in the right direction. The more everyday Skylines are offered with six-cylinder in-line engines ranging from 2.0 to 2.5 litres with rear or four-wheel-drive.

Best All-Rounder: Skyline GT-R

BODY STYLES: Coupe, Saloon
ENGINE CAPACITY: 2.0, 2.5, 2.5 Turbo
PRICE FROM: n/a
MANUFACTURED IN: Japan

Nissan Rasheen

With an appearance which owes more to the Blue Peter school of cardboard and glue than to the styling studios of one of the world's major car manufacturers, the Rasheen turns heads if nothing else. Originally a special one-off motor show car, it was so well received in Japan that production became a certainty. With simple boxy styling and very few curves, it certainly is a different approach to four-wheel-drive.

The Rasheen is not as big as it appears, for it is based around the mechanical package of the old Nissan Sunny estate, which was sold as a 4x4 in the domestic market. The result is an engine of just 1.5-litres which has a hard task in giving the heavy Rasheen decent performance on the motorway, particularly as automatic transmission is standard. The interior is very car-like, but neither the ride nor the noise levels are likely to help long-term comfort. The bull bar, roof rails and huge sunroof are standard features.

Best All-Rounder: Rasheen 1.5

BODY STYLES: Estate
ENGINE CAPACITY: 1.5
PRICE FROM: n/a
MANUFACTURED IN: Japan

Oldsmobile Cutlass

Sister to the new Chevy Malibu, the Camry-sized Oldsmobile Cutlass aims a little higher in its call to Middle America. New for '97, the Cutlass offers sound build quality, a well-equipped and spacious cabin and a no-demands driving experience for not a lot of dollars. Styling is fairly conventional - it's vaguely Nissan Maxima-ish - with smooth flanks and a rather bland face. The interior also has that efficient, if somewhat sterile, Japanese feel.

Oldsmobile engineers wisely spent their time stiffening-up the Cutlass's bodyshell. That means less commotion from under the bonnet and a smoother, more refined ride. But the rest of the Cutlass driving experience is commuter-efficient and undemanding, but certainly pleasant enough. The standard twin-cam 2.4 four-cylinder is lively but rather too vocal. The optional 3.1 V6 is more appealing in that it offers extra torque and more subdued noise levels.

Best All-Rounder: Cutlass V6

BODY STYLES: Saloon
ENGINE CAPACITY: 3.1V6
PRICE FROM: $17,000
MANUFACTURED IN: US

Oldsmobile Bravada

Luxury four-by-fours are selling faster in the US than fireworks on the Fourth of July. Now Oldsmobile has jumped on the bandwagon with its own spiffed-up, off-roader. Based on the cheaper Chevrolet Blazer, the Olds Bravada comes with more leather than a Harley-Davidson convention, wood trim and standard fixtures like air con, power everything and even a compass. Digital, of course. With its Land Rover Discovery price tag, it's not cheap, however.

Refinement is the Bravada's big selling point. Its standard 190 bhp 4.3-litre V6 is well-muted, its four-speed automatic swaps ratios with the driver hardly noticing and, most importantly, it rides as smoothly as many an upscale saloon. Few buyers will ever venture off-road but if they do, permanent four-wheel drive is one of the Bravada's many standard features. Here, a computer decides where to direct the power depending on which wheel is gripping.

Best All-Rounder: Bravada

BODY STYLES: Estate
ENGINE CAPACITY: 4.3V6
PRICE FROM: $29,500
MANUFACTURED IN: US

Peugeot 106/**Citroen** Saxo

Two cars with different names and noses but basically the same under the skin. Citroen and Peugeot are both part of the giant French PSA group, and this 'new' car is in reality a heavy development of the old 106. The changes add 11cm to the length, a stronger body and improved safety features. Power steering and airbags are available for the first time, while the engines are either new or modified with a high performance 1.6 106/Saxo available from 1997.

The result is an easy-to-drive, solid-feeling small car which offers high levels of ride comfort, smoothing out even the bumpiest of roads, yet has handling which is fun for keen drivers too. The popular engine choices are a bit disappointing though - the 1.1 is a little slow and, together with the 1.4, can be noisy except when cruising at a steady speed. The front seats will be very comfortable for most people, but space behind is tight compared with much of the competition.

Best All-Rounder: Saxo 1.1 SX

BODY STYLES: Hatchback
ENGINE CAPACITY: 1.0, 1.1, 1.4, 1.6, 1.5D
PRICE FROM: £7,150
MANUFACTURED IN: France

Peugeot 306

The Peugeot 306 sets the standard for refinement and driver appeal in the Escort class and the wide range of models means there's something for just about everyone. The usual 1.4, 1.6 and 1.8 engines are the mainstay, but there are also some of the best diesels in the business as well as a sporty 2.0. And new from the middle of 1996 is the GTI-6, a 167 bhp 2.0-litre with a six-speed gearbox. The body style doesn't end with the pretty hatchback. You could also choose a striking convertible or a more conservative four-door saloon .

One thing they all share is a double helping of driver appeal, with great ride and handling the 306 trademark. Inside the seats prove comfortable but rear space isn't generous, headroom is compromised if a sunroof is fitted and the dashboard lacks a quality touch. The bread-and-butter petrol-engined models are in some ways overshadowed by both the fine turbo-diesel versions and the sporty 306 XSi, but in truth there's not a duffer in the range.

Best All-rounder: 306 XRdt

BODY STYLES:	Hatchback, Saloon, Convertible	PRICE FROM:	£10,100
ENGINE CAPACITY:	1.4, 1.6, 1.8, 2.0, 1.9D, 1.9TD	MANUFACTURED IN:	France, UK

Peugeot 406

On sale for just a year, the 406 has already established itself firmly as a strong contender in the family-car stakes. A stylish exterior makes it one of the most elegant family saloons; in keeping with tradition Peugeot have ignored the hatchback option, preferring an estate (launched in late 1996) and, for 1997, a 406 coupe. The engine range is expanding steadily, to include 1.8, 2.0 and 2.0 petrol turbo, plus 1.9 and 2.1 turbo-diesels.

In some areas the 406 competes well with executive cars costing thousands more. The interior is understated but very classy and spacious - although headroom is tight in cars with the optional sunroof. As with the 405 it replaced, the steering and handling is of a very high order, though comfort is compromised a little by a firmer ride and front seats which need more lumbar support. None of the engines disappoint in terms of performance, though the 1.8 is rather vocal.

Best All-Rounder: 406 2.0 LX

BODY STYLES:	Saloon, Estate, Coupe	PRICE FROM:	£12,800
ENGINE CAPACITY:	1.6, 1.8, 2.0, 2.0 Turbo, 1.9TD, 2.1TD	MANUFACTURED IN:	France

Peugeot 605

Peugeot's executive-saloon 605 has never really caught on in most markets, despite popularity in its native France. In Britain, its 'grown-up 405' appearance lacks the cachet of German rivals able to flaunt a glossier image, while the performance has never really impressed. That's not only because the V6 and 2.0-litre petrol engines are undistinguished, but also because the best of the turbo-diesels, the 2.5, is unavailable in the UK.

Those who try the 605 will discover a car with many hidden talents. Not least of these is a cosseting ride quality that sponges up B-road bumps as well as it levels motorway undulations. Light controls and a sumptuously trimmed, well specified and commodious interior makes the 605 comfortable and refined but, it must be said, the interior lacks class. Of the engines on offer, the 3.0 V6 appears under powered and none-too-smooth, which leaves the 2.0-litre four-cylinder or the surprisingly refined 2.1 TD as the pick of the range

Best buy: Peugeot 605 SRTD

BODY STYLES:	Saloon	PRICE FROM:	£21,500
ENGINE CAPACITY:	2.0, 2.0 Turbo, 3.0V6, 2.1TD	MANUFACTURED IN:	France

Plymouth Prowler

Chrysler's funky hot rod for the Nineties hits the showrooms next January. But put away your cheque-book, it's already sold out for a year. Probably three. Retro looks hide bang-up-to-date technology. An aluminium chassis, a 3.5-litre V6 that sounds like a V8 and an 'Auto-Stic' four-speed manual/auto transmission, with the gearbox at the back. Okay, so you can only get it in purple, but it looks great. And there's no luggage space, so buy the optional trailer and tow it behind.

Cars just don't come any cooler than this. Arm resting on the door, sun-shades on, just listening to the burble of that V6. Awesome. Yes, it would be more purist to have a throbby V8 up front but the 214 bhp 'six' from Chrysler's Concorde does just fine. You won't light-up the rear tyres but when you have 295/40s on 20 inch rims, nothing's going to unstick them. More than anything, the Prowler is a blast to drive.

Best All-Rounder: Prowler + trailer

BODY STYLES: Convertible
ENGINE CAPACITY: 3.5V6

PRICE FROM: $38,000
MANUFACTURED IN: US

Plymouth Breeze

Value for money doesn't get much better than this. A new Plymouth Breeze, with standard air conditioning, power steering, cruise control and a punchy, 2.0-litre 16-valve engine, sells Stateside for the equivalent of just under £10,000. It's the bargain basement version of Chrysler's mid-size four-door, which sells in Europe as the Stratus. In addition to its low price, the Breeze boasts vast interior space, an athletic chassis and a fun, breezy character.

The 2.0-litre engine is the same as that used in Chrysler's perky Neon. With 132 bhp on tap, it makes for lively, enthusiastic performance, particularly when mated to the standard five-speed shifter. Laser-accurate steering and a chassis that is responsive and well-balanced, yet is adept at soaking up the bumps, gives the Breeze a fun, youthful character. Add to this, an interior almost as spacious as a Ford Scorpio's and this new Plymouth makes a great buy.

Best All-Rounder: Breeze

BODY STYLES: Saloon
ENGINE CAPACITY: 2.0

PRICE FROM: $14,500
MANUFACTURED IN: US

Pontiac Grand Prix

Two years ago, Pontiac revealed a swoopy, in-your-face concept car called the GPX. They even whispered that the next Grand Prix would look a lot like the GPX. No one believed them. Well, the '97 Grand Prix looks almost identical to the GPX. Just as bold. Just as in-your-face. Available in four-door and two-door coupe bodies, the Grand Prix gets its power from either a lame 3.1-litre V6, a punchy 3.8-litre 'six' or an optional 240 bhp supercharged V6.

If you want your Grand Prix with a heavy dose of excitement, you go straight for the blower. The supercharged Grand Prix GTP packs its best punch from standstill up to 60 mph, getting there in well under seven seconds. Despite all that power going through the front wheels, the steering is free from wayward tendencies and is decently accurate. It rides well and the standard auto 'box changes smoothly and when you want it to.

Best All-Rounder: Grand Prix 3800

BODY STYLES: Saloon, Coupe
ENGINE CAPACITY: 3.1V6, 3.8V6, 3.8V6 Supercharged

PRICE FROM: $19,300
MANUFACTURED IN: US

Pontiac Sunfire

Americans pay the equivalent of £8,000 for a nicely-equipped Pontiac Sunfire coupe. Fifty quid more buys them the smart four-door version. For that, they get a striking-looking Escort-sized four-seater, with air conditioning, standard dual airbags and anti-lock brakes. There's even a fun convertible version with power top and a glass rear screen, as well as a sporty GT coupe with big wheels and a punchy 2.4-litre twin cam.

Of course, you pay for what you get and in the Sunfire's case, it's a gruff-sounding 2.2-litre base engine and even gruffer twin-cam. But performance with both is lively - the 2.4 is positively rapid - and even the base model has a fun, chuckable feel. Inside, there's space for four, though the rear seat is best-suited to kids. Build quality is a little iffy, with flimsy-feeling plastic trim and a few squeaks and rattles. But it's still great value.

Best All-Rounder: Sunfire SE 2-door

BODY STYLES: Saloon, Coupe
ENGINE CAPACITY: 2.2, 2.4
PRICE FROM: $11,500
MANUFACTURED IN: US

Pontiac Bonneville

Imagine a Scorpio with some serious muscle and you're looking at Pontiac's supercharged Bonneville. Restyled a year ago to get shot of some of the glitzy chrome and add a more distinctive nose and tail, Pontiac's flagship is an ample five-seater that's loaded to the gunwales with neat gizmos. How about the jet fighter-like head-up display on the windscreen, or a Computer Command Handling Pack? Fun stuff.

The Bonneville gets its poke from an engine that's essentially been around since 1962. But there's nothing ancient about the Bonneville's 240 bhp supercharged 3.8-litre V6. Squeeze the throttle and the engine releases its 280 lb ft of torque with the same immediacy as flicking a light switch. It'll hit 60 mph in seven seconds and the ton in 19.7. That's quick. Downsides? That Magnasteer steering provides plenty of tactile information, but has a hard job of taming the motor's torque on take off. The brakes need beefing-up too.

Best All-Rounder: Bonneville SE

BODY STYLES: Saloon
ENGINE CAPACITY: 3.8, 3.8 Supercharged
PRICE FROM: $22,000
MANUFACTURED IN: US

Porsche Boxster

The first all-new Porsche in almost 20 years, the Boxster is a cause of delight among enthusiasts everywhere. Unlike the 911 with its rear engine, and the front engined 968 and 928 (both dropped in the last year), the Boxster is mid-engined to give the perfect weight distribution and therefore, in theory at least, the best possible handling. Like the 911, the engine is a flat six, but this time it's water rather than air-cooled. With 2.5 litres and 204 bhp, Boxster will accelerate to 60 mph in 6.9 seconds and reach a top speed of 149 mph. A five-speed Tiptronic automatic gearbox is an option.

Porsche intends the Boxster to be as practical as it is fast. The short engine allows plenty of space for two passengers and their luggage - there's a boot at the front and rear which will even take golf bags. The electric roof will open or close in a record 10 seconds and there's a purpose-built hard top for the winter.

Best All-Rounder: Too soon to say

BODY STYLES: Convertible
ENGINE CAPACITY: 2.5
PRICE FROM: approx. £30,000
MANUFACTURED IN: Germany

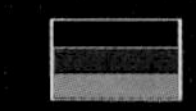

Porsche 911

Now within the last year of its extraordinarily long life, the 911 still somehow manages to remain up there on top of most drivers' wish list. And rather than wind down the range, Porsche keeps adding new models and refinements to maintain the image. Latest is the Targa, a name that has been used in the past for the 911 with a removable roof panel, but now it refers to the all-glass roof which cleverly slides back behind the rear window. The engine power has also been increased a touch, to 285 bhp for the standard 3.6-litre models which are available with two or four-wheel-drive. Fastest 911 is the 4wd Turbo, but drivers can get the looks if not the gut-wrenching performance with the lookalike Carrera 4 S.

The 911 is now more user-friendly than ever. No longer an especially difficult car to drive (which will disappoint some), it's comfortableand rides well. But it remains a highly idiosyncratic car, all the better for its wailing engine and simple interior design.

Best All-Rounder: 911 Carrera

BODY STYLES: Coupe, Convertible, Targa
ENGINE CAPACITY: 3.6, 3.6 Turbo
PRICE FROM: £61,000
MANUFACTURED IN: Germany

Proton Compact

With a new Mitsubishi Colt hitting the streets in 1996, Proton took over the old model. The similarity is apparent but none the worse for that, for the Compact is one of the prettiest hatchbacks around. Mechanically it is close to the Persona and the story is the same inside, with the facia straight from the bigger car. Indeed it was sold as the Persona Compact before Proton needed to conjure up another model after the MPi range faded away.

Initially, the Compact feels pretty good. It is well finished, easy to drive and comes with a great three-year warranty. Three engines, 1.3, 1.5 and 1.6, offer respectable levels of performance, though they, and the suspension, lack refinement. But the key question which keeps surfacing concerns its size and price. Although it's longer than any supermini, the space for rear passengers and their luggage is actually inferior to most. That means it's not quite the bargain it first appears.

Best All-Rounder: Compact 1.5 GLSi

BODY STYLES: Hatchback
ENGINE CAPACITY: 1.3, 1.5, 1.6
PRICE FROM: £9,000
MANUFACTURED IN: Malaysia

Proton Persona

If the original Mitsubishi Lancer-based Proton's MPi model was out of date even before it hit European shores, its newer and larger stable mate, the Persona, is a much more convincing effort. Offered in two versions, saloon and five-door hatchback, the Persona competes head-on with the likes of the Ford Escort. A facelift in 1996 saw safety improvements - a driver's airbag - and the option of a 1.8-litre petrol or 2.0-litre diesel added to the existing 1.5 and 1.6 engines.

On the road, most versions produce decent performance, particularly the 1.8 which is very lively. The diesel, however, is let down by excessive noise. Power steering is now standard and the easy-to-drive nature makes the Persona painless to use. There are a few quibbles; the trim looks drab in some versions, the rear passenger space is at a premium and the ride can be joggly around town. In Britain, perhaps the most important reason for choosing one is the extensive warranty package.

Best All-Rounder: Persona 1.6 XLi

BODY STYLES: Hatchback, Saloon
ENGINE CAPACITY: 1.5, 1.6, 1.8, 2.0D
PRICE FROM: £10,700
MANUFACTURED IN: Malaysia

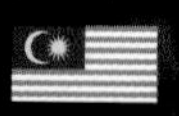

Renault Twingo

A runaway success in its homeland, the characterful Twingo looks set to acquire the mantle of the legendary Citroen 2CV for its blend of chic style and practicality, and all for a budget price. The concept is a simple one - offer only one version of the car, with a single engine and trim, to keep prices in check. The policy has been diluted a little, with air conditioning, an electric pack and, for '97, electric power steering, as options, but the idea still works.

1997 also sees the Phase 3 Twingo, with a fresh range of colours and Renault's new 1.2-litre engine. On the road it rides with the self-assured composure that only small cars of Gallic origin seem able to achieve, and well able to maintain a decent pace on the motorway. Four or five adults can be accommodated without feeling crushed but there's not much of a boot beneath the luggage hatch. The bad news is it's only available in Britain by special import – and it comes only in left-hand drive.

Best All-rounder: Twingo 1.2

BODY STYLES: Hatchback
ENGINE CAPACITY: 1.2
PRICE FROM: approx. £7,000
MANUFACTURED IN: France

Renault Clio

Renault's cheeky Clio received a facelift for 1996, the second major improvement since its launch five years earlier. A new bonnet, rounder headlights and new bumpers are the key exterior features, but there are inevitable safety improvements too along with a much-needed new 1.2-litre engine. Otherwise the Clio range is slightly reduced, with the high performance 16-valve and Williams versions dropped, leaving just two more petrol engines - 1.4 and 1.8 - plus a 1.9 diesel.

The new 1.2 may not have 16-valves like some competitors, but it provides smooth, fuss-free performance and easy motorway cruising. The other engines are not so quiet, though improvements to sound deadening should help. Comfort levels are reasonable in a sort of squidgy French-car way, though others better the Clio for rear legroom. Perhaps the biggest disappointment is the ugly, disorganised dashboard

Best All-Rounder: Clio 1.4 RT

BODY STYLES: Hatchback
ENGINE CAPACITY: 1.2, 1.4, 1.8, 1.9D
PRICE FROM: £7,500
MANUFACTURED IN: France

Renault Megane

You could be forgiven for thinking Renault was going for world domination with its new Megane. No less than seven body styles will be available when the range is completed in 1997, aimed at capturing buyers from across the spectrum. Big seller will be the five -door hatchback, but there's also a coupe, saloon, estate, convertible and van. Most interesting is the Scenic, a sort of mini Espace with two-rows of seats which can be adjusted or removed to suit your needs.

The Megane is a well thought-out car, with much attention to detail. It's comfortable, there's plenty of room, the interior is very pleasing and it's easy to drive. Critically, however, it's not much fun to drive, in the way that a Golf or Escort can be. Soggy steering, body roll on corners and a soft driver's seat put paid to that, while even the 1.6 engine doesn't feel particularly brisk. .

Best All-Rounder: Megane 1.6 RT

BODY STYLES: H'back, Saloon, Estate, C'vertible, MPV, Coupe
ENGINE CAPACITY: 1.4, 1.6, 2.0, 1.9D, 1.9TD
PRICE FROM: £11,700
MANUFACTURED IN: France

Renault Laguna

It has been a long time since Renault achieved distinction with a family car, but with the Laguna it has a model which competes on even terms with the best. Introduced in 1994, the Laguna started life as a five-door hatchback but the range was extended in 1996 to include an estate which is available in five or seven-seat form - the Family. The engine range has rapidly been extended since launch to form one of the broadest available in car of this type - 1.8, 2.0, 2.0-16-valve, 3.0-litre V6 plus a 2.2 diesel and turbo-diesel.

The Laguna is one of the classiest family cars around. The detailing outside is neat, while inside the interior and facia are solid and distinctive. Seat comfort and a good ride have long been a Renault strength, though those in the back of the car are less well-served. The Laguna is good to drive, though not in an overtly sporty way - few of the engines provides the ultimate urge of the better competitors. But it is a refined and distinctive car.

Best All-Rounder: Laguna 2.0 RT

BODY STYLES:	Hatchback, Estate
ENGINE CAPACITY:	1.8, 2.0, 3.0V6, 2.2D
PRICE FROM:	£11,700
MANUFACTURED IN:	France

Renault Safrane

Renault has long had a problem with its Safrane. Successful as it may be on its home soil, few buyers outside of France take French executive-class cars seriously. There's no denying that it's a very different animal to its German counterparts, yet the Safrane appears to hold many of the virtues expected of a top-level express.

It's a soft-riding car, comfortable and spacious, with deeply upholstered seats and featherweight steering. In all forms it's well equipped and it handles with assurance. But it has lacked both the image and the solid feel so necessary to executive car buyers, while none of the engines inspire . For 1997 the Safrane gets a comprehensive facelift, restyled both inside and out. Significantly, it also has some new engines, the 2.0-litre 16-valve from the Laguna, a 2.5 five-cylinder from the Volvo 850 and, though not for the UK, a 2.2 turbo diesel. It could be enough to bring the Safrane into favour.

Best All-Rounder: Too soon to say

BODY STYLES:	Hatchback
ENGINE CAPACITY:	2.0, 2.5, 3.0V6, 2.2TD
PRICE FROM:	approx. £17,500
MANUFACTURED IN:	France

Renault Espace

The current Renault Espace pictured right is now in its last throws, with a new model starting to hit European markets by late 1996. It's a replacement that is sorely needed, for after 11 years of total domination of sales of MPVs in Europe, the Espace has been bettered by numerous newcomers. Most notably cars like the Ford Galaxy, Citroen Synergy and their close cousins offer a better driving experience - the Espace's pedals are particularly awkward and the engines uninspired.

The new Espace retains the composite body construction, dimensions and general external style of the old car. Inside, however, it is radically different, with the facia dominated by a central speedometer - other controls are on the steering-wheel or doors. The rear seats no longer have fixed positions but are mounted on rails which run the length of the deck, allowing for great flexibility. It could be enough to put the Espace back on top.

Best All-Rounder: Espace T' Diesel

BODY STYLES:	Multi Purpose Vehicle
ENGINE CAPACITY:	2.0, 2.2, 2.9V6, 2.1TD
PRICE FROM:	£16,800
MANUFACTURED IN:	France

Renault Spider

1996 became the year of the sports car, with only BMW failing to make the break in getting right-hand-drive versions of its Z3 to Europe. But while Alfa, BMW and MG represent the sophisticated end of the market, the Renault Spider, along with the Lotus Elise, is something very different, a bare bones sportster where comfort and luxury take a distinctly second place to the goals of performance and handling. The most obvious result of this policy is the use of an air deflector in place of a windscreen, although a more traditional screen is offered on right-hand-driver Spiders.

The technologically-advanced chassis is built from aluminium, the bodywork glass-fibre and the mid-mounted engine taken from the Clio Williams. The result is a chassis which provides phenomenal levels of grip and handling, but an engine which is not really powerful enough to exploit the potential. Still, the Spider grabs attention like few other cars, and gives drivers a thrill a minute.

Best All-Rounder: Spider 2.0

BODY STYLES: Convertible
ENGINE CAPACITY: 2.0
PRICE FROM: approx. £25,000
MANUFACTURED IN: France

Rolls-Royce

Change happens at a measured pace at Rolls-Royce, so the arrival of a new engine and the reintroduction of an old name rank as important events. As with Bentley, the Rolls-Royce Silver Spur benefits from light turbocharging for its 6.7-litre V8 engine, increasing the power by 25% to 300 bhp and offering a serious improvement in performance. The new Silver Dawn is a replacement for the Silver Spirit, and with the emphasis much more on luxury than with the Silver Spur, it retains the normally-aspirated V8.

While there will always be those who argue that Mercedes-Benz makes a better luxury car, they miss the point. The fact is, the Rolls has an ambience and British charm that others can't hope to capture. The values that make a Rolls special are the refinement, the interior quality, the prestige. The feather-soft ride together with a big V8 engine and its almost imperceptible auto-shift ensures it is easy to cruise along with grace.

Best All-Rounder: Silver Spur

BODY STYLES: Saloon
ENGINE CAPACITY: 6.7, 6.7 Turbo
PRICE FROM: £119,000
MANUFACTURED IN: UK

Rover Mini

1997 sees a major repositioning of the Mini in Rover's line-up. Its move upmarket had already begun - in '96 it cost £1,300 more than the modern Fiat Cinquecento - but now Rover wants the Mini to be considered more as a fashion statement than as a cheap runabout. Two models run side-by-side, the Mini and the Mini-Cooper, with a starting price of around £8,500 for either. A wide range of option packs - retro, luxury, sports - allows the buyer to customise the appearance of their car. In the extreme, Cooper buyers can opt for a chequered roof, fully chromed fitments and 13-inch alloy wheels.

All models now take what was the 63 bhp Cooper engine, which has been revised to come with the latest emission regulations. There is still the old four-speed gearbox, although the radiator has been moved to the front of the car. Safety is much improved, with door beams and a driver's airbag.

Best All-Rounder: Mini Cooper

BODY STYLES: Saloon
ENGINE CAPACITY: 1.3
PRICE FROM: approx. £8,500
MANUFACTURED IN: UK

Rover 100

With its origins going back to the original Metro of the early 1980s, the Rover 100 is a surprisingly competent package. 'Roverisation' brought a new grille, bumper and headlamps to give the 100 the family look. The front seats were narrowed to improve the feeling of spaciousness and there were detail changes to the switches and facia. The engines are the familiar 1.1 and 1.4-litre K-Series, with a 1.5 diesel from Peugeot. Alongside the three and five-door hatchbacks runs an expensive convertible.

An old design it may be, but the 100 manages to project a feeling of quality unavailable in most small cars. Comfort levels for those in the front are high, while the ride and noise levels are quite acceptable. The petrol engines are very refined, with even the 1.1 giving sprightly performance. The Metro feels small and nippy around town, but the downside it that it really is small - space for rear passengers is about as cramped as you'll find in any supermini.

Best All-Rounder: Rover 111

BODY STYLES: Hatchback, Convertible
ENGINE CAPACITY: 1.1, 1.4, 1.5D
PRICE FROM: £7,200
MANUFACTURED IN: UK

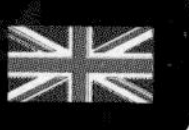

Rover Niche

With the introduction of the new 200 hatchback in 1996, a ragbag of models evolving from the old 200 and 400 ranges remains. The Coupe, Cabriolet and Tourer estate all survive, each subject to some tinkering and rejuvenation. All the petrol engines were replaced by the impressive K Series found in the newer 200/400, either 1.6 or the 1.8 VVC from the MGF (though demand from MG buyers has postponed introduction of the 1.8 to 1997). Continuously variable automatic transmission also became an option for the first time.

Each of these related cars has an elegant interior, with stainless steel plates on the sills and wood on the doors and dashboard. Comfort levels are good, with fine seats and ride, although room in the rear of the Coupe and Cabriolet is tight. The Tourer has a limited floor area and is more of a stylish estate than a heavy-duty load carrier; it is the only model to come with the option of a Peugeot turbo diesel engine.

Best All-Rounder: Tourer 1.6

BODY STYLES: Coupe, Convertible, Estate
ENGINE CAPACITY: 1.6, 1.8, 2.0TD
PRICE FROM: £16,000
MANUFACTURED IN: UK

Rover 200

Now becoming established as the company which offers something a little different to mainstream rivals, Rover's new 200 places more emphasis on style and classy fittings than on building the roomiest car in its class. For while the 200 competes on price with hatchbacks like the Astra, it is several inches shorter. Engines range from 1.4 to the 2.0-litre VVC unit from the MGF, as well as two turbo-diesels.

It is very much a car of great strengths and weaknesses. Both inside and out the design can't help but be admired, with even the most basic version getting bits of wood and stainless steel and comfortable well-trimmed cloth seats. They all perform well too, no matter what the engine. The downside is that most engines are too noisy when much is demanded from them, the ride will be too firm for some and the rear seats are only suited to shorter adults.

Best All-Rounder: 216 Si

BODY STYLES: Hatchback
ENGINE CAPACITY: 1.4, 1.6, 1.8, 2.0D
PRICE FROM: £11,000
MANUFACTURED IN: UK

Rover 400

When Rover launched the 400 in 1995, it insisted the hatchback should be considered alongside cars like the Mondeo and Cavalier on the grounds that, although the 400 was clearly Escort-sized, the quality of the fittings and finish was a class apart. No matter how kindly you look at Rover that's a little hard to swallow, but things have changed with the arrival of the 400 saloon. Significantly longer than the hatch, it might conceivably be put alongside other mainstream medium-sized family cars.

Based on the Honda Civic five-door, the Rover gets its own K-Series engines, grille and interior treatment. There are plenty of the nice details which make Rovers seem special and the overall levels of comfort and refinement are impressive for a car of this size, although rear seat accommodation is rather poor. The original 1.4 or 1.6-litre range has been extended with a 2.0-litre and two very economical turbo-diesels. All are a delight to drive.

Best All-Rounder: 416 Si saloon

BODY STYLES: Hatchback, Saloon
ENGINE CAPACITY: 1.4, 1.6, 2.0, 2.0TD
PRICE FROM: £13,000
MANUFACTURED IN: UK

Rover 600

Rover's compact executive car, the 600, offers buyers the opportunity to combine classy styling with a great sense of British occasion. For while the exterior inevitably owes much to the Honda Accord from which the 600 was developed, the interior has all those nice touches so beloved of Rover owners - strips of wood, chrome and stainless steel make it all seem rather special.

Those in the front have plenty of the room - the seats are particularly large - but, like the BMW 3 Series and Audi A4, strong competitors, room in the rear isn't generous. A good new 1.8-litre variant was introduced in '96 which overshadows the existing 2.0-litre. Both of these are Honda engines, like the 2.3, although Rover offers its own punchy turbo-diesel and the very powerful turbocharged 620ti. Each engine has much to offer, though the balance on all 600s is towards comfort rather than the taught sportiness of a BMW. It's a pity, then, that the ride lets it down, too firm except on the smoothest roads.

Best All-Rounder: 618 Si

BODY STYLES: Saloon
ENGINE CAPACITY: 1.8, 2.0, 2.0 Turbo, 2.0TD
PRICE FROM: £15,500
MANUFACTURED IN: UK

Rover 800

There was a time when Rover's 800 was the best selling executive car in the UK. But no more. Time and other manufacturers have caught up the with large Rover. Improvements continue to be made though, lately a new Rover-designed 2.5-litre V6 to replace the Honda unit and changes to the suspension to improve the handling. All derivatives are available as a hatchback or saloon, with elegance the key, the smart grille and tasteful 'wood and chrome' interior still managing to make the 800 seem a cut above many other executive cars. As well as the V6, Rover offers 2.0-litre engines, in standard and turbocharged form, as well as a turbo-diesel.

The 800 isn't as roomy as many competitors, the lack of width making it tight for three in the back. The mainstream models are lively and drive well enough, though they will never excite. The Vitesse has uprated suspension and better steering to make the whole experience much more enjoyable.

Best All-Rounder: 820 SLi

BODY STYLES: Saloon, Hatchback, Coupe
ENGINE CAPACITY: 2.0, 2.0 Turbo, 2.5V6, 2.5TD
PRICE FROM: £18,500
MANUFACTURED IN: UK

Saab 900

Saab's smaller saloon manages to convey an air of being rather special. First, it's very solidly built. Then there are the safety features which are unsurpassed in this price range and a host of unique Saab touches - an ignition lock on reverse gears rather than by the steering wheel, the (optional) Sensonic clutchless gearchange and the 'black panel' which allows all but essential instruments to be turned off. There's a choice of three body styles, the mainstream five-door hatch, a three-door hatch which Saab optimistically refers to as the Coupe, and a convertible.

The 900 is a comfortable car though in the rear it's not particularly roomy with the large front seats restricting the view forwards. Every engine, from the 2.0-litre through to the 2.3 and 2.0 turbo to the 2.5 V6, provides eager performance, but the gearchange and lack of handling precision lets it down as a true driver's car. 1997 sees the launch of an even higher performance 900.

Best All-Rounder: 900 2.0i XS

BODY STYLES: Hatchback, Coupe
ENGINE CAPACITY: 2.0, 2.0 Turbo, 2.3, 2.,5V6
PRICE FROM: £14,400
MANUFACTURED IN: Sweden/Finland

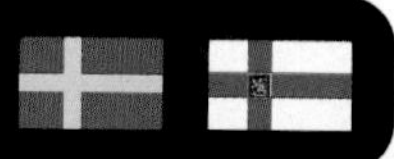

Saab 9000

Having become something of an old stager, the 9000 range is in its last days, with an all-new model due for launch in the spring of 1997. That will be a CD saloon, but the current CS hatchbacks will run for at least a year after that. The existing range major on Saab's speciality, turbocharging, though lately the emphasis has been redirected away from the ultimate performance models towards the 'Eco' turbos which provide more thrust at lower speeds. Also available are non-turbo engines, a 2.0-litre four-cylinder unit and a 3.0 V6.

The 9000 is a solid executive car which provides comfortable room - and safety - for five adults. The interior can be a bit oppressive, with huge front seats and an imposing dashboard, which the cheaper models do little to alleviate with too much dull plastic and fabric. The 2.0-litre drives well enough but the Eco models, with their willing performance, are the most impressive.

Best All-Rounder: 9000 2.0 Eco

BODY STYLES: Hatchback, Saloon
ENGINE CAPACITY: 2.0, 2.0 Turbo, 2.3Turbo, 3.0V6
PRICE FROM: £18,400
MANUFACTURED IN: Sweden

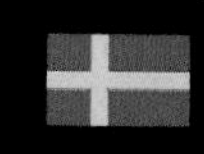

Saturn SC

After re-shaping its Astra-sized saloon and estate last year, Saturn has now got around to updating its pretty two-door coupe for '97. The underpinnings haven't changed but the new body is a little more striking and harder-edged, and the freshened interior finally gets dual airbags. As before, two versions are on offer; the SC1 with its 100 bhp single-cam 1.9-litre, and the more spirited 124 bhp 1.9 twin-cam. Saturn's dirt-cheap pricing applies to the new coupe.

As before the new coupe looks better than it goes. The problem? Buzzy, vocal engines that go into spasms of protestation when they're revved hard. If you don't mind the noise, then the twin cam coupe is a lively performer, delivering a solid thump in the back away from traffic lights or out of corners. The chassis keeps up the sporty appearance too, having nimble steering, a firm, well-controlled ride and an overall feeling of agility.

Best All-Rounder: Saturn SC2

BODY STYLES: Coupe
ENGINE CAPACITY: 1.9
PRICE FROM: $10,500
MANUFACTURED IN: US

Seat Marbella

The ubiquitous holiday hire car, the box-shaped Marbella is something of an anachronism in a world which has moved on to softly-curved styling. The design, a licensed copy of Fiat's Panda, harks back to the Seventies – with little further development since. Cheap and basic transport was the brief, and it's certainly that. Although the Marbella is no longer on sale in the UK, it's still available in many European countries.

There's no denying the Marbella has a certain unorthodox charm though there's little else to recommend it. The ride can be tortuous, the gearchange is sloppy, the steering is vague and involves far too much arm-twirling when negotiating roundabouts and the handling is decidedly odd. Inside, the archaic facia and flimsy switchgear hark back to another age, there's precious little in the way of boot or rear passenger space, and the build and trim are flimsy with no safety features to speak of. No wonder it's no longer on sale here.

Best All-Rounder: The cheapest

BODY STYLES: Hatchback
ENGINE CAPACITY: 0.9

PRICE FROM: n/a
MANUFACTURED IN: Spain

Seat Ibiza/Cordoba

This VW-owned Spanish car-maker has put a massive effort into improving the quality, safety and overall design of its cars, while retaining some Spanish flair. Nowhere is that more clearly illustrated than with the latest Ibiza supermini and its close cousin, the Cordoba saloon. Both rely heavily on VW-sourced components and engines while actually sharing the main platform and dashboard with VW's award-winning Polo, which can't be a bad thing.

The Ibiza has a solid feel and spacious interior. The range covers three and five-door hatchbacks with engines from a puny 1.0-litre to a potent 1.8-litre 16-valve. Plenty of others, including diesels, fill the gaps between, including a new sporty 100 bhp 1.6 in the two-door Cordoba SX, which Seat would like us to think of as a coupe. On the road, both the Ibiza and Cordoba are thoroughly capable cars. They feel tightly screwed together and are pleasant inside, but a slightly harsh ride spoils the comfort to some extent.

Best All-Rounder: Ibiza 1.4 Salsa

BODY STYLES: Hatchback, Saloon
ENGINE CAPACITY: 1.0, 1.4, 1.6, 1.8, 2.0, 1.9D, 1.9TD

PRICE FROM: £7,500
MANUFACTURED IN: Spain

Seat Toledo

Seat's answer to the likes of the Mondeo and Vectra is its Toledo. At one time this VW-engined family car sold largely on price alone, but it has matured over the years and, in its latest form, it looks convincing and comes with a hearty equipment and safety specification. The Toledo remains great value; even the base models benefit from power steering, twin airbags, electric sunroof, air conditioning and central locking. VW-sourced engines are of 1.6, 1.8 and 2.0-litre capacity but there's also a sporting 16v 2.0 and no less than three 1.9 diesels.

Although it looks like a saloon, the Toledo is a hatchback and a spacious one at that. Rear legroom and boot space are among the best in the class. VW-ownership for Seat means that quality is so much better in the latest models but there's still the odd trim creak. Ride comfort lags behind some of the newer cars but the Toledo handles with a pleasing crispness.

Best All-Rounder: Toledo 1.8 SE

BODY STYLES: Hatchback
ENGINE CAPACITY: 1.6, 1.8, 2.0, 1.9D, 1.9TD

PRICE FROM: £12,100
MANUFACTURED IN: Spain

Skoda Felicia

Since Volkswagen bought Skoda in 1991 the Czech manufacturer's only model has matured very nicely. Now benefiting from the options of a VW 1.6 petrol engine and (soon) a 1.9 diesel, it is still the original 1.3 that keeps the Favorit within the price band where its merits really stand out. The range starts at just over £6,000 for the five-door hatchback, with estates £1,000 more. There is no automatic transmission but power steering is on the way.

The Felicia is a surprising package. The interior fittings and facia are of a much higher quality than the old Favorit while outside, although the heritage is obvious, the appearance has been tastefully uplifted. To drive the Felicia is never going to be an inspiring proposition, with even the more powerful version of the 1.3-litre engine providing flat performance - acceleration in fifth gear is notably weak. But as a cheap, simple, five-door runabout it's a fair package, with ample room for four adults and their luggage.

Best All-Rounder: Felicia 1.3 LXi

BODY STYLES:	Hatchback, Estate	PRICE FROM:	£6,200
ENGINE CAPACITY:	1.3, 1.6, 1.9D	MANUFACTURED IN:	Czechoslovakia

Spectre R24

If you are looking for a car as arresting as the McLaren F1 for little more than one tenth of the cost, the Spectre R42 could be your car. It's hi-tech all right, with a composite body-chassis unit which like much of the rest of the car owes more to racing car manufacturing methods than those traditionally associated with road cars. Heart of the Spectre is a mid-mounted Ford V8, its 4.6-litres producing 350 bhp which propels the car to 60 mph in 4 seconds and on to an electronically limited top speed of 175 mph.

The R42 is said to be as docile in traffic as it is wild on the autobahns. It's luxurious too, with leather racing-style seats and a walnut interior. Luggage space is confined to the low front compartment - 'long weekend needs' best describes its capacity. It's an interesting proposition, but it's not McLaren buyers the company should be worried about - at £70,000 it faces some heavy guns from the like of Porsche, Jaguar and Aston Martin

Best All-Rounder: Spectre R42 4.6

BODY STYLES:	Coupe	PRICE FROM:	£59,500
ENGINE CAPACITY:	4.6V8	MANUFACTURED IN:	UK

Ssangyong Musso

Ssangyong currently offers just one model for export - the four-wheel-drive Musso. Designed by Briton Ken Greenley and powered by a Mercedes-Benz diesel, the Musso is intended to combine the qualities of a car in an off-roader platform. Compared with most competitors, the Musso sits lower with road-car steering and suspension. Four-wheel-drive can be selected electrically on the move. A 220 bhp petrol-engined version is added in late 1996.

The space inside the Musso is enormous, with ample room for five six-footers and their luggage. It is comfortable too, with a well controlled ride and little of the severe body roll in corners normally associated with off-roaders. Noise levels are commendable when cruising, but the diesel can vibrate unpleasantly around town. The performance is distinctly pedestrian, although it cruises happily enough at motorway speeds.
Pricing is very sharp in the UK.

Best All-Rounder: Musso 2.9D SE

BODY STYLES:	Off-Roader	PRICE FROM:	£16,000
ENGINE CAPACITY:	2.3, 3.2, 2.3D, 2.9D	MANUFACTURED IN:	Korea

Subaru Impreza

Subaru's success in the world rally championship has had a profound effect on the sales of the Impreza, with around half of all buyers opting for the astonishingly quick 2.0-litre Turbo model. It's a bargain in anyone's book, offering Escort Cosworth-level performance for thousands less than the Ford. Most Imprezas come with four-wheel drive as standard, with the engine choice including 2.0 and 1.6. There's a choice of two body styles, a traditional four-door saloon, or a stylish five door which falls halfway between a hatchback and estate. The latest 2.0 Sport combines the Turbo's looks with the less powerful engine.

The Impreza comes across as a thoroughly likeable car to drive with an impressive ride. It handles well too, but both the non-turbo engines are best served with plenty of revs to give their best; a sluggish mid-range in the 1.6 spoils driveability. The Impreza's weakness is the deadly dull interior, with desperately plain dashboard and fabrics.

Best All-Rounder: Impreza 2.0 Sport

BODY STYLES:	Hatchback, Saloon	PRICE FROM:	£11,300
ENGINE CAPACITY:	1.6, 2.0, 2.0 Turbo	MANUFACTURED IN:	Japan

Subaru Legacy

The Legacy range of 4x4 estates and saloons has built up quite a following, which can only expand when US-built cars start to arrive in the UK soon, including the chunky-tyred, high ground-clearance Outlander version. For those living in hilly country where winter snows can make four-wheel drive a life-line to civilisation, the Legacy makes sense. It's useful for towing too, but it's a convincing enough car even without the need for this facility. .

Not least of these is a pliant ride and a strong and refined range of flat-four engines. The all-weather security of four-wheel drive is well worth having for many owners, and the overall high standard of build and finish marks this out as a car of quality. It's generally well-equipped and there's enough space for four without feeling cramped. Those looking for faults might point to the dreary trim, the estate's narrowish load bay and the heavy fuel consumption, but in all, the Legacy makes a versatile and well mannered family car.

Best All-Rounder: Legacy 2.0 GLS

BODY STYLES:	Saloon, Estate	PRICE FROM:	£13,700
ENGINE CAPACITY:	2.0, 2.0 Turbo, 2.2, 2.5	MANUFACTURED IN:	Japan, US

Subaru SVX

It may sell only in tiny numbers but Subaru's quirky SVX keeps on going year after year. From the all-alloy 3.3-litre flat-six engine to the glassy cockpit with its stylised side windows, Subaru's SVX defies convention while managing to create a swift and civilised cruiser. This is certainly more a tourer than a sportster and it carries all the expected accoutrements. Everything from leather trim and air conditioning to an auto transmission and cruise control are part of the SVX package.

Flat six engines can sometimes appear lumpy but the Subaru's 24-valve unit is quiet and smooth, delivering solid if not outstanding performance. It's rather softly-sprung but where it really scores is in its all-weather grip and high-speed refinement and comfort. Permanent four-wheel drive allows swift and safe progress in adverse conditions and the 226 bhp engine gives effortless cruising ability.

Best All-Rounder: SVX

BODY STYLES:	Coupe	PRICE FROM:	£30,500
ENGINE CAPACITY:	3.3	MANUFACTURED IN:	Japan

Suzuki Swift / **Subaru** Justy

While many car manufacturers are getting together to design and build new models, it is unusual for one to take on the completed design of another. That is what Subaru has done with its new Justy, taking the Hungarian built Suzuki Swift as its basis then adding four-wheel-drive and extra UK-applied anti-corrosion treatment.

Both are available with sporty three-door body styling, or as a longer five-door with improved rear legroom. The standard 1.3-litre engine provides an adequate blend of performance and economy and with the Suzuki there are two further options. The 1.0-litre three-cylinder engine is noisy when revved, as it needs to be to produce any meaningful urge, while the 100 bhp GTI model is a brisk and fun performer, if a rather raucous and harsh one. Recent design improvements add airbags and safety features but there is no escaping the fact that the Swift is now long in the tooth compared with mainstream European rivals.

Best All-Rounder: Justy 1.3 GX

BODY STYLES: Hatchback
ENGINE CAPACITY: 1.0, 1.3
PRICE FROM: £6,200
MANUFACTURED IN: Japan / Hungry

Suzuki X-90

Suzuki's cute little Cappuccino sports car may recently have ceased production but the X-90 is here to fill the fun-car gap. Based on the mechanics of the Vitara off-roader, the X-90 is a strict two-seater with a separate boot and lift-off glass roof panels. The familiar 1.6-litre petrol engine is used, with the option of either two or four-wheel-drive; automatic transmission is available as an option on the 4x4. Equipment levels are good, with twin airbags, power steering and central locking.

While the X-90 stands out in a crowd, with its distinctive appearance sporty, bucketed seats, it doesn't have much else going for it. The ride is bouncy and the handling remains pitched at off-road use rather than providing a sharp Tarmac drive. The performance is adequate with plenty of gear changing but there's the rub - the gearchange is really hard work. In the end, the X-90 just isn't half the fun its looks suggest.

Best All-Rounder: X-90

BODY STYLES: Off-Roader
ENGINE CAPACITY: 1.6
PRICE FROM: £10,000
MANUFACTURED IN: Japan

Suzuki Vitara

Suzuki has always been at the forefront of the leisure 4x4 boom with the Spanish-built Vitara model that really firing the imagination with its mini-Discovery appearance. The soft-top and three-door models continue to provide a fun car for the young-at-heart, but they are complemented by the Vitara estate, a roomy and practical alternative. More recently, the new V6 Vitara with its wider track, chunky new look and a silky 2.0-litre engine counters criticism of lack of pace and refinement, while an automatic transmission turbo-diesel at long last answers the needs of the economy driver.

These latest five-door models prove the Vitara has at last grown up. They ride far better than the choppy short wheelbase versions and have an understated look that appeals to a more mature buyer. However, the original three-door Vitara can't begin to compare with the likes of the excellent Toyota RAV4 for outright ability, ride or sophistication.

Best All-Rounder: Vitara V6

BODY STYLES: Off-Roader
ENGINE CAPACITY: 1.6. 2.0V6, 2.0TD
PRICE FROM: £11,900
MANUFACTURED IN: Spain

Suzuki Baleno

Suzuki's first mainstream family car was introduced in 1995. Available either as a four-door saloon or three-door hatchback, it is aimed at Escort-class cars, though it is clearly shorter than most of its competitors. For 1997 a five-door estate will go on sale in Europe. Airbags for both front occupants are a good safety feature and both the GLX saloon and GS hatchback are well-equipped, although a sunroof and stereo are options.

A 1.8 joins the range for 1997, though the 1.6 is already powerful enough to make the Baleno rocket off the line like a GTi. But it is an unrefined engine, unpleasantly noisy when driven quickly, while sadly the chassis doesn't maintain the sporty impression. Inside, the Baleno is well-finished with bucketed sports seats in the three-door, more subdued examples in the saloon. Space inside either isn't brilliant, but everything must be taken in context of the attractive pricing and great warranty.

Best All-Rounder: Baleno GS

BODY STYLES: Hatchback, Saloon, Estate
ENGINE CAPACITY: 1.3, 1.6, 1.8
PRICE FROM: £9,000
MANUFACTURED IN: Japan

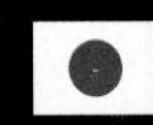

Toyota Starlet

While Toyota has undeniable success world-wide with its Corolla, it has never been totally convincing when it comes to superminis. The latest version of the Starlet was introduced in spring of 1996 and while it moves things along, it is still more of the same old formula. Available in three or five-door body style with just a 1.3-litre engine in the UK, the Starlet can also be bought in other countries as a diesel or with a punchy 1.3 turbo engine producing a gut-wrenching 135 bhp.

The standard 75 bhp engine is lively, pretty refined and produces good fuel economy. Like most Toyotas, it is easy to drive, with a slick gearchange and power steering on most versions. It isn't great fun however - the body heaves over too much on corners yet the suspension doesn't cope well with bumps either. The interior of the new car is a big improvement, but space for those in the rear and their luggage is merely average for this size of car.

Best All-Rounder: Starlet 1.3 CD

BODY STYLES: Hatchback
ENGINE CAPACITY: 1.3
PRICE FROM: £7,600
MANUFACTURED IN: Japan

Toyota Corolla

Toyota's small family car has for many years worn the crown of the 'best selling car in the world'. There is a huge range of models made in factories all over the world, including a saloon, estate, five-door 'liftback' and shorter three-door hatchback for the UK market as well as a high performance coupe in Japan. Engines range from 1.3 litres through diesels to high output 2.0-litre twin-cam models, though in the UK the top engine is a 1.6.

The success of the Corolla is down to doing most things well even if it comes top of the class nowhere apart from, perhaps, reliability. The front seats are firm but comfortable, the ride is acceptable, noise levels are kept in check and even the 1.3 drives well. The chief criticisms centre on the amount of interior space available - especially the three-door - and the fact that, as efficient and reliable it may be, the Corolla is a pretty dull car.

Best All-Rounder: Corolla 1.3 GS

BODY STYLES: Hatchback, Saloon
ENGINE CAPACITY: 1.3, 1.6, 2.0D
PRICE FROM: £10,000
MANUFACTURED IN: Japan

Toyota Carina E

Toyota's British built Carina E was subject to some changes for 1996, with improved equipment levels, including driver and passenger airbags across the range and a halving of the huge numbers of models on offer. As a competitor in the important fleet market for the Mondeo and Vectra, the Carina is offered in three body styles - saloon, hatchback and estate - although the emphasis in the revised range is very much on the hatch. Four engines are on offer, with 1.6 and 1.8 high-economy 'lean-burn' engines, 2.0-litre and a new 1.8 turbo-diesel.

In terms of comfort the Carina E scores highly. The seats provide good levels of support, the car rides well and noise levels are generally low. The space inside is among the best in the family car class and the same goes for the boot. The interior of many version is too uninspiring, however, with a plain dashboard and dull seat fabrics. It's an easy car to drive, though only the 2.0-litre gives much in the way of urge.

Best All-Rounder: Carina 2.0 GLi

BODY STYLES: Hatchback, Saloon, Estate
ENGINE CAPACITY: 1.6, 1.8, 2.0, 2.0TD
PRICE FROM: £11,800
MANUFACTURED IN: UK

Toyota Camry

Toyota has been churning out new models recently, with six hitting the showrooms during 1996. Latest is the fourth-generation Camry executive model. Now offered exclusively as a saloon, the Camry is slightly bigger in all key dimensions - length, width and height - than the outgoing model. Significantly it also has a longer wheelbase which increases the already generous amount of passenger space available. The seats have been redesigned to provide improved leg and lateral support, and there's more adjustment.

As before, two engines are on offer. The existing 2.2-litre is carried over, with either manual or automatic transmission. The all-aluminium 3.0-litre V6 is new, offering more power (199 bhp) yet improved fuel economy too. The V6 is coupled to a computer-controlled automatic transmission which varies gear changes according to your driving style. Naturally equipment levels are high, with twin airbags and fully automatic air conditioning standard.

Best All-Rounder: Too soon to say

BODY STYLES: Saloon
ENGINE CAPACITY: 2.2, 3.0V6
PRICE FROM: n/a
MANUFACTURED IN: Japan

Toyota Avalon

This is the Toyota from Kentucky. Built in America to compete with traditional American Yank Tanks, the Avalon has more legroom than any Japanese car ever sold in the US, including the Lexus LS400. It beats most American full-size cars too. And that's amazing considering it's based on a stretched Camry platform. Power comes from the Camry's 192 bhp, 3.0-litre V6 coupled to a slushy four-speed automatic.

The drawback to the Avalon's size is its narrow waistline. So, while Toyota claims there's room for six inside - it even offers a front bench seat - it's actually better-suited for four. The Camry-sourced V6 provides relaxed and near-silent performance, with particularly strong mid-range response and reasonable economy. Softish suspension means there's some body lean on curves but it provides a cushy ride on America's largely arrow-straight motorways. Think of it as a cheap Lexus.

Best All-Rounder: Avalon XLS

BODY STYLES: Saloon
ENGINE CAPACITY: 3.0V6
PRICE FROM: $24,000
MANUFACTURED IN: US

Toyota Paseo

The number of budget coupes on offer is expanding rapidly, with manufacturers trying to latch onto the success of the Tigra and Civic coupes. Latest is the Toyota Paseo, a front-wheel-drive hatchback with a single 1.5-litre engine on offer, an apparently advanced double-overhead camshaft 16-valve unit which produces a disappointing 89 bhp. Equipment levels are good, though, with power steering, sunroof, twin airbags and safety standards built around the higher 1998 standards.

The complete Paseo package is summed up by its styling - a safe, no-risks package. The exterior will offend no-one, but neither will it excite. The same goes for the interior which, apart from the white instrument dials could have come from any family Toyota. The Paseo is an easy drive, but hardly a quick car. It is comfortable enough, although headroom under the standard sunroof isn't generous; the back seats suitable only for small children.

Best All-Rounder: Paseo 1.5 ST

BODY STYLES: Coupe
ENGINE CAPACITY: 1.5
PRICE FROM: £12,500
MANUFACTURED IN: Japan

Toyota MR2

Just as a whole host of new sports cars flood onto the market, the last rites are being given for Toyota's successful MR2. With a year or so to run, the MR2 gets some final tweaks and safety improvements, and a slight drop in power to 168 bhp after adjustments to cope with new emission regulations. Like the MGF, but predating it by 10 years, the MR2 is mid-engined, with the fixed-roof GT complemented by the T-Bar with removable top panels.

It takes but a few hundred yards behind the wheel to appreciate that this is a proper sports car. That much is clear from the direct steering, the taut suspension, the nifty gearchange and the vocal 2.0-litre engine. The MR2 handles in a secure but entertaining fashion – unlike the tail-happy earlier versions – and the engine gives promising performance. Inside, it's hard to find anywhere to put keys or a wallet but the boot, set behind the engine, makes this a more practical car than might be expected.

Best Buy: MR2 GT

BODY STYLES: Coupe
ENGINE CAPACITY: 2.0, 2.0 Turbo
PRICE FROM: £21,000
MANUFACTURED IN: Japan

Toyota Celica

Unbelievably, Toyota has built more versions of its Celica than Ford has the Escort. In its latest incarnation - the sixth introduced late in 1993 - the development continues along the formula path of front-wheel-drive 2+2 coupe based around a 2.0-litre engine. Its power - 173 bhp - is impressive, though the engine needs working hard to extract the best from it. In addition there's a pretty cabriolet as well as the rally-car inspired GT-Four, with turbocharging and four-wheel-drive.

Behind the distinctive facade, the Celica is just a bit too predictable . The interior is the main culprit, which seemingly tries hard to be different but never really succeeds. There are many good points, however, notably the comfortable seats and a slight improvement in the level of rear seat room. It's sporty too, with eager performance and steering and suspension to match. The convertible is particularly impressive for its lack of buffeting with the roof down.

Best All-Rounder: Celica 2.0 GT

BODY STYLES: Coupe, Convertible
ENGINE CAPACITY: 1.8, 2.0, 2.0 Turbo
PRICE FROM: £17,700
MANUFACTURED IN: Japan

Toyota Picnic

Just when you thought no-one could dream up another niche for a new car, Toyota claims to have done it The Picnic is 'a new motoring concept, a Family Fun Vehicle'. Versatility and fun are top of the list of characteristics offered by the six-seat Picnic, with seats that fold, slide, recline or even completely disappear. There's easy walk-through access between all three rows of seats.

Already on sale in Japan as the Ipsum, where it is available with four-wheel-drive, the Picnic reaches Europe in the autumn of '96 in front-wheel-drive only. While the body structure and styling are certainly all new, the concept appears to owe quite a lot to the Mitsubishi Space Wagon, while mechanically the chassis is based upon that of the Carina E, with the 2.0-litre engine borrowed from the RAV4 off-roader. Still, there are strong possibilities that the whole idea could be a big success, even bearing the silly name in mind.

Best All-Rounder: Picnic 2.0

BODY STYLES:	Estate	PRICE FROM:	n/a
ENGINE CAPACITY:	2.0	MANUFACTURED IN:	Japan

Toyota Previa

Few of the current range of people carriers can match the Previa for carrying capacity, with only the forthcoming Chrysler Voyager looking like providing strong competition. The key to its success is in being able to carry seven or eight people and their luggage at the same time, although its futuristic styling also makes the Previa out as something special. Practicality is marred a little, however, by a sliding side door on one side only and by seating versatility that is limited compared with many of its European rivals.

On the open road, the Previa carries its bulk well and the high seating position is a boon for both driver and passengers. The 2.5-litre engine, although rather boomy, produces adequate pace and the ride is comfortable. However, some may find the Previa a handful in town where its bulk is hard to conceal. Excellent build, a strong image and masses of space mean the Previa should continue to fare well despite a glut of newcomers.

Best All-Rounder: Previa GL

BODY STYLES:	Multi Purpose Vehicle	PRICE FROM:	£18,800
ENGINE CAPACITY:	2.4	MANUFACTURED IN:	Japan

Toyota RAV4

Toyota pulled off a smart move when it introduced the RAV4 in 1994. Until then four-wheel-drive vehicles were hardly any fun to drive on tarmac. The RAV4 changed all that, with an emphasis on enjoyable road driving as well as doing pretty well on the slippery stuff too. The basis is a stylish, chunky body with a sparky 2.0-litre engine. With 133 bhp on hand, the RAV4 keeps up with hot hatches while the suspension is sophisticated enough to prevent all that body sway on corners so familiar in other off-roaders.

Three-door models are available in two specs, with the better GX getting alloy wheels and two removable alloy sunroofs. Space for rear occupants and luggage is cramped, however, a problem answered by the more recent long wheelbase RAV4 five-door. As a machine for dealing with difficult 4x4 conditions, the RAV4 is less successful, having neither the tyres nor a low-range set of gears. But for most owners, that won't really matter.

Best All-Rounder: RAV4 GX

BODY STYLES:	Estate	PRICE FROM:	£13,300
ENGINE CAPACITY:	2.0	MANUFACTURED IN:	Japan

Toyota Landcruiser Colorado

This all-new Landcruiser slots in beneath the gigantic Landcruiser '80' model and replaces the Toyota 4Runner. Available in either three-door short wheelbase or longer five-door body styles, there are standard width and 'wide-body' Colorados, with pumped up wheel arches and side body mouldings. Power for the majority is provided by a 3.0-litre turbo-diesel, with a 3.4-litre V6 petrol engine reserved for the top five-door VX. Four-wheel drive is permanently engaged and automatic transmission is an option.

The Colorado's somewhat anonymous exterior appearance is continued through to the interior, which is efficient but lacking in character. Five-door versions get three rows of seats, but to claim, as Toyota do, it is an eight seater is stretching things a bit. But you'll be more convinced by the way the Colorado drives. The diesel is quiet and refined once you get moving and puts in a good performance on the motorway or the rough stuff.

Best All-Rounder: Colorado 3.0 GX

BODY STYLES: Estate
ENGINE CAPACITY: 3.0TD, 3.4V6
PRICE FROM: £20,000
MANUFACTURED IN: Japan

Toyota Landcruiser

Many people are drawn to off-roaders because of their size. If that's the case, then they'll love the mammoth proportions of the Landcruiser VX. A vast 4.5-litre engine helps to convey its bulk with surprising gusto, or there's a 4.2 turbo-diesel with colossal torque. Up to seven people can be accommodated within its ample dimensions, with space left for a proper luggage bay. A smaller three-door Landcruiser is also on sale with 3.0-litre turbo diesel power.

The best aspect of driving a Landcruiser VX is that everyone keeps out of your way – even London taxis. In other respects, the Landcruiser is the next best thing to a Range Rover. It's smooth and powerful, it gobbles up the motorway miles or it will cross deserts with the best of them. Few things count against it if you can handle the size, though it lacks the Range Rover's air-suspended control through the turns, the interior isn't the most stylish and, petrol or diesel,it consumes fuel with relish.

Best All-Rounder: Landcruiser VX D

BODY STYLES: Estate
ENGINE CAPACITY: 3.0TD, 4.4, 4.2TD
PRICE FROM: £30,200
MANUFACTURED IN: Japan

TVR Cerbera

Like most successful manufacturers of two-seater, traditional sports cars, TVR decided it was time to make a move upmarket. The Cerbera is the result, a meshing of 2+2 coupe practicality with the V8 muscle that has become the trademark of a TVR in the 1990s. But this V8 is TVR's very own design, producing 350 bhp from its 4.2 litres. The Cerbera is packed full of novel features too, from the electric door release under the wing mirrors to instruments viewed through the upper and lower steering wheel spokes.

As a performance car there is little to beat the Cerbera. It howls angrily off the line, accelerating faster than a Porsche 911 Turbo to top out at over 180 mph. Thankfully there's handling to match, and a surprisingly compliant ride. As a practical coupe, however, the Cerbera is less of a success, with an impossible cramped rear seat and no spare wheel, just a can of puncture sealer.

Best All-Rounder: Cerbera 4.2

BODY STYLES: Coupe
ENGINE CAPACITY: 4.2V8
PRICE FROM: £37,000
MANUFACTURED IN: UK

TVR Chimaera

The Chimaera may be humbled by the more brutal Griffith in terms of outright stomp and aggression, but in many ways it emerges as the better car to drive. The balance between its ample 4.0 or 5.0-litre V8 urge and grip from the tyres is a far more acceptable one for those of us with conventional driving skills, while the chassis balance is more appropriate for a car that spends its life on the road rather than a track. Yet there's still a vast amount of performance and soul in this great-looking car.

The interior is beautifully finished in leather and wood, the seats are comfortable and the Chimaera rides well for such a sporting machine. It's not a car for the faint-hearted though. The gearchange is very heavy, as is the steering without the optional power assistance. The low seating position reduces the visibility and, for all its apparent grip, drivers need care and skill to keep things in check in wet conditions.

Best All-Rounder: Chimaera 4.0

BODY STYLES: Convertible
ENGINE CAPACITY: 4.0V8, 5.0V8
PRICE FROM: £29,500
MANUFACTURED IN: UK

TVR Griffith

As hairy-chested British sports cars go, there's little to challenge the follically abundant Griffith. The bulges and scoops tell their own story of race-bred engineering. Heart of it all is a Rover-derived V8, developed by TVR to produce a gut-wrenching 340bhp – enough for 160mph. The space-frame chassis that forms the backbone of the car is almost an art-form in itself, but it does a thorough job of containing the excesses of power to viable proportions.

Sitting in the Trad-Brit cockpit is like being at the helm of a fighter-plane, only far more comfortable and set off with lashings of hand-trimmed leather. This is a man's car in every sense. The gearchange and clutch are heavy, the steering demands a firm grasp and a skilled one. It handles well enough up to a point, but anyone expecting five litres in a light car to behave like a pussycat will be in for a shock. It bites. Best reserved for those who like their meat raw.

Best All-Rounder: Griffith 5.0

BODY STYLES: Convertible
ENGINE CAPACITY: 5.0V8
PRICE FROM: £34,500
MANUFACTURED IN: UK

Vauxhall/Opel Corsa

The Corsa supermini offers a clever option of two distinct body styles based on the same basic structure. The three-door has a sporty swept tail while the five-door is more upright which allows for increased space in the back. Both pack in a great number of safety features, although not all of them are on the standard equipment list; options feature heavily when you buy a Corsa. A wide range of engines is available, though these have recently been cut back to just three petrol and two diesel in the UK.

For a car of this size the Corsa offers reasonable levels of room, comfortable seats and a good ride. The dashboard is very clear and the driving position and controls well thought out. Few Corsas score highly in the fun-to-drive stakes, however. The steering doesn't give a great deal of feel, the handling is not very crisp and the gearchange is often notchy. But it's safe and well built, valuable virtues in this class of car.

Best All-Rounder: Corsa 1.2

BODY STYLES: Hatchback
ENGINE CAPACITY: 1.2, 1.4, 1.6, 1.5TD, 1.7D
PRICE FROM: £7,400
MANUFACTURED IN: Spain

Vauxhall/Opel Astra

The Astra is now well past its mid-life crisis and remains one of the biggest selling small family cars in Europe. That's helped by a huge model range which encompasses three- and five-door hatchbacks as well as a saloon, estate and convertible. Four petrol engines and a couple of diesels make the permutations mind-boggling, though Vauxhall blows hot and cold about keeping a performance model in the range.

The quality immediately impresses, the Astra feeling a very solid and well-built car. The interior is well finished and comfortable, with adequate room for four adults. The 1.7 low-pressure diesel and the 1.6-valve engines provide excellent performance in their own way and although the 1.4 goes well too, it is somewhat noisy and unrefined. The worst point about the Astra, however, is the gearchange which can be stiff and reluctant to engage first gear. Equipment levels are good even on the cheapest Merit model.

Best All-Rounder: Astra 1.7D Merit

BODY STYLES: Hatchback, Saloon, Estate, Convertible
ENGINE CAPACITY: 1.4, 1.6, 1.8, 2.0, 1.7D, 1.7TD
PRICE FROM: £10,500
MANUFACTURED IN: UK/ Belgium

Vauxhall/Opel Vectra

Vauxhall carried over many of the styling details which characterised the Cavalier to the Vectra, so much so that from some angles the two cars can be confused. They shouldn't be, for the Vectra is largely new, with just the engines and transmissions carried over from the Cavalier. As before, saloons and hatchbacks are available, with an estate arriving in late 1996. Five petrol engines are offered and for '97 Vauxhall has a whole new range of turbo-diesels, including a 2.0-16 valve direct injection unit for the Vectra.

The Vectra has been criticised for being too conservative and it certainly isn't a car to set the pulse racing. It is more clinically efficient than the Mondeo or Peugeot 406, with the result that drivers will be impressed with solid build and quality of the switches, less so with the beauty of the car. To be fair, the Vectra does the job, providing good levels of room and comfort and an enjoyable drive. It just fails to be the standard setter anticipated.

Best All-Rounder: Vectra 1.8 GLS

BODY STYLES: Hatchback, Saloon, Estate
ENGINE CAPACITY: 1.6, 1.8, 2.0, 2.5V6, 2.0TD
PRICE FROM: £12,900
MANUFACTURED IN: UK/ Germany

Vauxhall/Opel Omega

Vauxhall's (and Opel's) Omega is as close as you'll get to the feel of a BMW or Mercedes at the price of a routine executive car. Stylish and solid, early cars haven't been entirely fault free, but the overall impression remains of a comfortable car with excellent safety features and far more equipment than you would expect from the two German rivals. There's a choice of two 2.0-litre and two V6 engines as well as BMW's six-cylinder turbo-diesel The range of trim options for the saloon and estate version is huge.

The Omega offers high standards of room and comfort for the occupants yet few compromises to its on-road ability - this a large saloon that drivers can enjoy too. The 2.5-litre V6 is beautifully smooth yet economical too, while the other strong choice is the turbo-diesel. The balance of attributes in the Omega seems well-judged except that it lacks that indefinable something that makes it truly desirable rather than just a good car.

Best All-Rounder: Omega 2.5 V6 GLS

BODY STYLES: Saloon, Estate
ENGINE CAPACITY: 2.0, 2.5TD, 2.5V6, 3.0V6
PRICE FROM: £18,250
MANUFACTURED IN: Germany

Vauxhall/Opel Tigra

Like or loathe the Tigra's styling, there's no denying it makes a strong statement. Based on the chassis of the Corsa supermini, the Tigra started life as one of those motor show styling exercises which thankfully General Motors had the courage to bring into full production. Tell-tale signs of the Corsa are easy to spot on the facia, but the Tigra is none the worse for that. The advantage of the link between the two cars is that cost is kept in check - the Tigra is good value for money.

But above all it's fun. There's a choice of two engines, with even the smaller 1.4 giving gutsy performance and a rorty bellow when driven hard. Bucketed front seats keep those in the front comfortable but interior space is pretty compromised. Small adults can just fit in the back seats but in front the position of the pedals makes the driving position somewhat cramped. The Tigra's practical, though, with a fair-sized boot and low running costs.

Best All-Rounder: Tigra 1.4i-16v

BODY STYLES: Coupe
ENGINE CAPACITY: 1.4, 1.6
PRICE FROM: £12,000
MANUFACTURED IN: Spain

Vauxhall/Opel Calibra

The Calibra has been, more than any other car, responsible for the resurgence of coupe sales in the UK. With its high-waist styling and tiny headlights, the Calibra has always been an extremely distinctive car, even though beneath the glitz there's only a Cavalier with a new body. That is particularly evident inside, where the dashboard has been lifted straight from the family car. The cleverest aspect, however, is the packaging. This is the only coupe which will sit four adults in relative comfort

Sales success is helped by a seemingly never-ending succession of special editions, usually based on the least powerful engine but with nice alloy wheels and a leather interior. For a car with sporting pretensions, however, the handling isn't really taught enough, although there is a good range of engines to compensate. 2.0-litre models are available with either 8 or 16 valves, with more relaxed performance provided by the 2.5-litre V6.

Best All-Rounder: Calibra 2.0 16v

BODY STYLES: Coupe
ENGINE CAPACITY: 2.0, 2.5V6, 2.0Turbo
PRICE FROM: £18,700
MANUFACTURED IN: Germany

Vauxhall/Opel Sintra

Vauxhall is rather late in the game with its first multi-purpose vehicle, the Sintra, following on from a flush of new models introduced in 1994. Designed to compete with MPVs on both sides of the Atlantic, the Sintra is naturally bigger than much of the European built competition, and claims to offer seating for eight people and their baggage. It's not too big, however, just 200 mm longer than a Vectra, so manoeuvring will not be too awe-inspiring. Sliding rear side doors ensure easy access to the second and third rows of seats.

Two engine options are offered, a 2.2 litre with 141 bhp and a 3.0-litre V6 producing a mighty 201 bhp, making this potentially the fastest MPV you can buy. A turbo-diesel follows in mid-1997. For a big boxy vehicle the Sintra is surprisingly slippery through the air, which should help the fuel consumption and noise levels.. Vauxhall has stated that all versions will have twin airbags, air conditioning, electric windows and alloy wheels.

Best All-Rounder: Too soon to say.

BODY STYLES: Multi Purpose Vehicle
ENGINE CAPACITY: 2.2, 3.0V6
PRICE FROM: n/a
MANUFACTURED IN: US

Vauxhall/Opel Frontera

Claimed to be Europe's best selling recreational vehicle, the Frontera succeeds through a combination of rugged, no-nonsense styling and value for money. Based upon a Isuzu design, but manufactured in Luton, the Frontera is available in a short-wheelbase 'Sport' version as well as a much roomier long wheelbase five-door estate. The range was revised a couple of years ago with redesigned suspension and a fresh range of engines; for 1997 there's yet another new diesel, this time a 2.5-litre direct-injection unit.

The Frontera is a much improved vehicle, with the performance, particularly of the heavier five-door version, more acceptable with either type of engine. The ride is improved too and though it still falls short of the better off-roaders, overall comfort levels are reasonably good. At least they are in the five-door; the shorter Sport models are much less user-friendly, with little comfort for rear passengers and not much in the way of cargo space either.

Best All-Rounder: 2.5 TDS Estate

BODY STYLES: Off Roader
ENGINE CAPACITY: 2.0, 2.2, 2.5TD
PRICE FROM: £15,000
MANUFACTURED IN: UK

Vauxhall Monterey/ Isuzu Trooper

Isuzu had already been selling the latest Trooper for two years in the UK before Vauxhall started importing the Monterey from Japan in 1974. Badge and trim levels apart, these cars are identical. Available either in short wheelbase three-door form, or as a long wheelbase five-door, the Monterey/Trooper is pitched headlong into competition with the best-selling Land Rover Discovery.

It's one of the bigger off roaders, particularly in its long-wheelbase form. That makes it very spacious inside and it's comfortable too, with the option of a fold-away sixth and seventh seat in the boot of the five-door. But there's little to entice you into the Monterey/Trooper in the same way that a Discovery is so appealing - the dashboard looks more like it came from a truck than a car. Both engine options - 3.2 V6 or 3.1 turbo-diesel - are great, however, and despite its size the car is surprisingly straightforward and easy to drive on the road.

Best All-Rounder: 3.1 TD

BODY STYLES: Off Roader
ENGINE CAPACITY: 3.1TD, 3.2V6
PRICE FROM: £18,000
MANUFACTURED IN: Japan

Venturi 400

Venturi builds the French equivalent to the Lotus Esprit. Like the Lotus, the Venturi has a turbo-charged engine mounted behind the front seats, but there the similarity ends. The engine is derived from a Renault V6, with a capacity of 2.8 or 3.0 litres, and power varying from 260 bhp to 400 bhp. The composite bodywork covers a complex semi-monocoque steel chassis.

Even in basic form, as the 260, the Venturi is very rapid indeed, with communicative steering and a chassis to match the high performance. The 2.0-seater interior is beautifully appointed, with acres of leather and walnut, so it's a comfortable car even when you're not going fast - apart from the engine, which never sounds especially nice. Top of the luxury range is the Atlantique 300, while those after outright performance choose the 400 GT, a detuned version of the Le Mans car. In addition to the coupe, there's a cabriolet with a neat rigid roof.

Best All-Rounder: Venturi 260

BODY STYLES: Coupe, Convertible
ENGINE CAPACITY: 2.5, 2.8, 3.0 - all V6Turbo
PRICE FROM: approx. £65,000
MANUFACTURED IN: France

Volkswagen Polo

VW's offensive on the supermini class with its latest Polo continues to shake up the opposition. It competes strongly on all fronts with a car which rivals the pacesetters in almost every area to make this arguably the best in its class. The wide range of engines covers the usual 1.0-litre to 1.6-litres, but there's also a 1.9 diesel with catalytic converter – one of the cleanest running engines of its type.

The first thing that strikes when driving a Polo is it feels like a bigger car. Not only does it sit firmly on the road but it also feels solidly constructed and well finished too. It's quiet and refined, it has a capable chassis and it's easy to drive. A good specification with some excellent safety features adds further to its charm. No faults then? Well, the rear passenger space is bettered by some rivals, and the ride and seats are rather firm. And prior to 1997, the 1.0-litre was desperately slow - now it has a little more power.

Best All-Rounder: Polo 1.4 L

BODY STYLES: Hatchback, Saloon
ENGINE CAPACITY: 1.0, 1.4, 1.6, 1.9D
PRICE FROM: £7,400
MANUFACTURED IN: Spain

Volkswagen Golf and Vento

There's something for just about everyone in the Golf and Vento line-up. Whether it's a hatchback or saloon, a convertible, a hot-hatch or an estate, it can be found here. This is all complemented by a wide range of engines from clean-running catalysed diesels through to a high-performance V6. One thing they all share is a tough and durable build that has become a hallmark of Volkswagen. They also share, in most cases, a certain lack of dynamism although there are some great engines - the new 1.4, plus the 2.8 V6 and TDi.

As a long-term prospect, it's hard to beat a Golf for the way its interior and fittings seem to shrug off the miles. The panels seem far less prone to dents than French rivals, and the feeling of newness remains longer than most, although neither the ride nor the handling are anything like as good. For sporting appeal, the Golf VR6 is exceptional. The TDi offers exceptional economy combined with good performance.

Best All-Rounder: Golf 1.4

BODY STYLES: Hatchback, Saloon, Convertible
ENGINE CAPACITY: 1.0, 1.4, 1.6, 1.9D
PRICE FROM: £10,250
MANUFACTURED IN: Germany

Volkswagen Passat

A brand new Passat arrived late in 1996 to replace the thoroughly worthy but ultimately dull VW range topper. Already one of the roomiest cars in its class, the new Passat is bigger still. The interior is said to provide a 'hint of sporting flair' coupled to very high safety standards - twin airbags and anti-lock brakes come as standard, with side airbags, built into the front seats, available as an option. Even the cheapest Passat comes with power steering, central locking and electric front windows.

Power is provided by a wide range of petrol and diesel engines, most of which are familiar from the Audi A4. The high performance V6 is available only with the Syncro (four-wheel-drive) transmission. While a traditional automatic transmission is available on lower-powered Passats, the performance cars will get the option of a 'tiptronic' five-speed automatic, which offers complete manual control if the driver chooses.

Best All-Rounder: Too soon to say

BODY STYLES: Saloon, Estate
ENGINE CAPACITY: 1.6, 1.8, 1.8 Turbo, 2.3V5, 2.8V6, 1.9TD
PRICE FROM: approx. £13,000
MANUFACTURED IN: Germany

Volkswagen Beetle

Yes, the lovable bug is still being churned out in Mexico and Brazil at the rate of 500 a day. Still readily identifiable as Ferdinand Porsche's original concept as a people's car for Germany, it's hard to believe it was actually designed pre-War. Almost unique in its front booted format with horizontally opposed air-cooled rear engine, few cars can offer quite so much character. It's a fact which is recognised by Volkswagen, which plans to introduce an all-new Beetle (see New for 1997) within the next year.

On the road, the Beetle's distinctive but weedy engine can be heard whistling away far behind, and although the ride and handling might barely be considered acceptable these days, it's unreasonable to expect anything else from a design this ancient. It will cruise happily at 70 mph, nonetheless, but don't expect a Beetle to tackle corners with any finesse or with much grip. The Beetle lives well in the past when it comes to safety features and durability.

Best All-Rounder: Beetle 1.6

BODY STYLES: Saloon
ENGINE CAPACITY: 1.6
PRICE FROM: approx. £6,000
MANUFACTURED IN: Mexico/ Brazil

Volvo 400

Volvo has tried to compensate for the age of its smallest car by packing the 440 hatchbacks and 460 saloons full of safety and other equipment to give it an edge. The cheapest 400s include driver's airbag, sunroof and power steering in their specification, from which five other variants can be derived - Family (S), Sport (Si), Business (SE), Performance (GLT) and Luxury (CD). Each can be matched up to any of the three petrol or turbo-diesel engine.

But despite all this effort, the Volvo 400 feels like an elderly car, both to sit in and to drive. The dashboard is heavy and intrusive and, while the front seats are comfortable enough, room in the rear isn't generous and the ride is unrefined. The engines provide reasonable performance but none compares with refinement achieved by more modern competitors. For safety and value for money the 400 takes some beating, but look elsewhere if you are after 1990s standards of comfort and driving enjoyment.

Best All-Rounder: 440 1.8 Si

BODY STYLES: Hatchback, Saloon
ENGINE CAPACITY: 1.6, 1.8, 2.0, 1.7 Turbo, 1.9TD
PRICE FROM: £11,600
MANUFACTURED IN: Netherlands

Volvo S40/ V40

A joint venture with Mitsubishi, and manufactured side-by-side in Holland, the S40/V40 represents a new stylish approach from Volvo, which naturally maintains the strong safety stance that has done the company so well for so long. The S40/V40 fits between the 400 and 850 range of Volvos, to compete with the likes of the Ford Mondeo and Vauxhall Vectra. The S40 is the saloon, V40 the 'lifestyle' estate, with more emphasis on style, less on function than with previous Volvo wagons.

Two engines are available, with the lively 2.0-litre much the best proposition - the 1.8 is noisier when revved hard, which it has to be at times. These are enjoyable cars to drive, particularly with the Sports suspension package - Volvo offers a huge number of options on the basic car. The interior is roomy in cars without a sunroof, while, oddly, seat comfort which is only average in front is great in the back. A car with a lot of appeal, especially as an estate.

Best All-Rounder: V40 2.0 SE

BODY STYLES: Hatchback, Estate
ENGINE CAPACITY: 1.8, 2.0
PRICE FROM: £14,200
MANUFACTURED IN: Netherlands

Volvo 850

Volvo's 850 set a new standard for this traditional Swedish car-maker by providing far greater driver appeal than any previous big Volvo. That it has a five-cylinder engine and front-wheel drive is another break with tradition, but it's the sporty T5 model with its 225 bhp engine and threateningly aggressive stance that really tells us Volvo is changing its image. The 850 proves it's possible to have traditional Volvo values of safety without it being dull.

With no less than seven engine options all between 2.0 and 2.5 litres, there is no shortage of choice and, with the addition of a coupe and four-wheel-drive estate for 1997, the options keep on growing. While the T5 provides storming performance it is the lesser engines which offer the best all-round balance, including the turbo-diesel from Audi. As you might expect, most 850s are sold in estate form where they have a reputation for being tough and commodious. An ugly facia and an awkward gearchange are the weak points.

Best All-Rounder: 850 2.5 10v SE

BODY STYLES: Saloon, Estate
ENGINE CAPACITY: 2.0, 2.3 Turbo, 2.5, 2.5TD
PRICE FROM: £18,000
MANUFACTURED IN: Sweden, Belgium

Volvo 940/ 960

These big Volvos never fail to polarise opinion. Loved by antique dealers, dog lovers and caravanners but loathed by trendies, the Swedish-built cars are made to last with a strong emphasis on safety, but the razor-edge styling and dowdy appearance dulls the appeal. The 960 is the more sophisticated model with its all-independent suspension and six-cylinder engines. The 940 remains only for those who require a cheaper big Volvo and find the 850 isn't quite large enough; it comes with either a four cylinder petrol or six-cylinder turbo-diesel engine.

In estate trim the 900-series has always impressed not just for its load swallowing ability, but for the safety structure and strong engines too. But mediocre ride and handling together with a dated interior reduces the appeal of the 940. A heavily revised but visually similar 'new' 960 arrived in 1994, providing much improved comfort and road manners.

Best All-Rounder: 960 2.5 S Estate

BODY STYLES: Saloon, Estate
ENGINE CAPACITY: 2.3 Turbo, 2.4TD, 2.5, 3.0
PRICE FROM: £16,500
MANUFACTURED IN: Sweden

Westfield

It's no accident that the Westfield has a marked similarity to the classic Lotus 7. But don't dismiss this as merely a fake; the engineering is top-level, tough enough to withstand engines up to high-power V8s. If the 300 bhp V8 version proves integrity, it's certainly sufficient for the likes of the Ford 16v engines that form the mainstay of the range. Most of these, like the Caterham 7, are built in kit form, but Westfield also supplies cars fully built.

Compared to the Caterham 7, it has a more tail-happy and marginally less composed stance through the corners, but this spells pure fun for the driver who likes to drive by the seat of his pants. Acceleration is dependent on engine choice but even the smaller engines provide vivid performance; with modified V8 power it's awesome. The cockpit is cramped, especially for the feet, and the wind buffeting at speed is a constant reminder that this is not for everyday driving.

Best All-Rounder: Westfield 1800

BODY STYLES: Convertible
ENGINE CAPACITY: 1.6, 1.8, 3.9V8
PRICE FROM: £11,800
MANUFACTURED IN: UK

ARO 4x4

We used to know the Aro 10 in Britain as the Dacia Duster. It never had much of a reputation and is no longer sold here but it remains in production in its native Romania. Based on Renault 12 mechanicals combined with four-wheel drive, the Aro 10 is the Eastern European equivalent of a Suzuki Samurai: small, cheap and useful off-road. A variety of body styles is offered, including soft and hard tops, short and long wheelbases, pick-ups and crew cabs. All models have free-wheel hubs, towbar, bull bar and twin headlamps. Two engines are offered: the old 1.4-litre Renault 12 engine or a Renault 1.9-litre diesel.

FSO Caro

After being wooed by GM and Fiat, Poland's FSO has in fact jumped into bed with Daewoo of Korea, which intends to invest over $1 billion over the next five years in new models. In the meantime, FSO plugs on with the Caro, essentially an 18-year-old design, itself based on the floorpan of the Fiat 125 of the 1960s. Nowadays the mechanical side at least is more up-to-date, with a choice of 1.5-litre petrol and 1.9-litre Citroen diesel engines. All now have power steering, Girling brakes and a better-looking dash. FSO will soon be using Rover's 1.6-litre K-series engine and the five-door hatchback may be joined by a four-door saloon and an estate.

Hindustan Ambassador

A relic of British colonial rule in India has returned home to Britain. Hindustan still builds the Ambassador - a 1950s Morris Oxford - under licence at the rate of 50,000 per year from works in Calcutta. These days it might have a catalysed 1.8-litre Isuzu engine and a five-speed gearbox but the rest of the specification is distinctly olde-worlde: leaf sprung rear and all-round drum brakes. You can even order an original wood veneer Morris dashboard, two-tone paint and leather upholstery for the 'Fifties look. Its low price and spaciousness attract about six UK customers a month.

Hindustan Contessa

Amazing as it may seem, this is India's luxury car. It will be instantly recognisable to many as a licence-built version of the 1972 Vauxhall Victor FE. Hindustan calls it the Contessa Classic and has updated it slightly by fitting glassfibre wrap-around bumpers and plastic wheel trims. Compared to other domestic products, it really does seem like a limousine. Like the Ambassador, it is fitted with an Isuzu 1.8-litre engine coupled with a five-speed gearbox. The Contessa also has something the old Victor never did - the option of air conditioning, which is almost essential in India. It is not available in the UK.

Mahindra CJ

These days the Mahindra is sold as an off-road vehicle with occasional on-road capability rather than the other way around. The reason is clear: the Mahindra is an extremely capable machine off the beaten track but very crude in everyday use. If the Mahindra looks rather like a World War II Jeep, it should come as little surprise, since this Bombay-built off-roader was originally licensed from Willys as long ago as 1949. Available in long and short wheelbase lengths, it is available with Peugeot 2.1-litre diesel or 1.9-litre engines. Extras include rear seats, winch, metal doors, alloy wheels and even carpets.

Moskvich 2141

Russia's attempt to come up with a modern family car in fact produced something which looks like an old Chrysler Alpine. The name Moskvich disappeared from Britain some 20 years ago and has never dared return but the more up-to-date 2141 was briefly marketed in Europe as the Aleko, without success. As well as a Moscow-built petrol engine the 2141 is also available with a Ford-derived diesel engine. It is a completely conventional front-wheel drive car sold as a five-door hatchback or a five-door estate. Production numbers at the Moscow factory have plummeted in recent years as the outdated designs face up to harsh economic realities.

Premier 118

India's protectionist economy allows the continued production of cars considered well past their sell-by date by western standards. The Premier 118NE is an old Fiat cast-off, the 124, also still made in Russia as the Lada Riva and in Turkey as the Tofas. Made in Bombay, the 118NE differs little from the original Fiat model except it is powered by a choice of Nissan petrol or diesel engines. By Indian standards it really is 'spacious, contemporary and dynamic' as the brochure claims and it is certainly a cut above the Hindustan. However, Premier has engineered a real coup - a licence to build the Peugeot 309!

Tata Ghurka

The name Tata has already endeared itself to many commercial users, attracted by a range of pick-ups and vans with very low prices. Now Tata has launched a passenger model in the UK whose name - Gurkha - hints at its Indian origins. Styling is a cross between Mercedes-Benz 190 and Vauxhall Frontera but the Gurkha is no 4x4. It's a rear-wheel drive only diesel estate car with a huge carrying capacity. Standard equipment includes power steering, electric windows and central locking and, on the SE version, alloy wheels, bull bar, roof rack and stereo. The best thing about the Gurkha is its price: from just £9,995, it's a real bargain.

Tatra 700

It's not often there's something new to report from Czech-based Tatra, the world's second oldest surviving marque. But the 700 is as new as it gets. It's a revised version of the old 613, first seen in 1970. The bodywork is substantially restyled and the interior has been modernised, but the floorpan and mechanical side are very much as before. A rear-mounted air-cooled V8 engine in a luxury car may sound off-putting, but in fact it delivers safe handling. Performance is excellent thanks to considerable British engineering input. Tatra hopes to boost production to 650 per year, and right-hand drive is available.

Tofas Sahin

Turkey's motor industry rotates around two companies: Otosan (a Ford subsidiary) and Tofas (Fiat). Both are modernising rapidly, Otosan now making the current Escort and Tofas the Tempra. Tofas also still makes a version of the old Fiat 131, a model first seen as long ago as 1974. In saloon form it's called the Sahin and as an estate it's the Kartal. Technically it hasn't changed much from the 1970s but it's been made to look more modern with plastic bumpers and wheel trims. Amazingly, this model is still popular in its birthplace, Italy, where a fair number are exported for conservative-minded customers.

VW Gol

'Gol' means goal in football-mad Brazil and Volkswagen's Brazilian subsidiary has certainly scored with this new car. It's Brazil's best-selling car (the old model sold over 1.5 million) and a genuine answer to the imported European hatches which are starting to gain a hold in an increasingly open Brazilian economy. Although its styling cues are taken from Wolfsburg, the whole car is Brazilian designed and made. As well as the base 1.0-litre, four fuel-injected engines are offered (1.0, 1.6, 1.8 and 2.0), each capable of taking petrol or alcohol fuel, plus a 1.6 diesel. There are hatch, saloon, estate and even GTI models.

ZAZ Tavria

With the break-up of the former Soviet Union, Russia's second largest car maker, ZAZ, has devolved to the Ukraine. The only model this firm now makes is the 1102 (or 1105 as a five-door), known in export markets as the Tavria, which has now taken over from the old rear-engined 968. It's a small front-wheel drive hatchback with a 1.1-litre or 1.3-litre engine. Its specification is extremely basic, which at least allows it to be sold for a bargain price: indeed, in those European countries where it is sold, it is the cheapest car you can buy. The Lada importers wanted to do the same in Britain but the project faltered early on.

Technical Data

The table on the following pages outlines the technical background to cars listed in the Guide. By and large we have used the car manufacturers' own figures, so in places where they are unable or unwilling to release the information we have placed a dash. This is usually because the car is new, although American car makers are notoriously reluctant to release performance data.

Engine capacity is given in cubic centimetres, power in bhp, although the common metric power unit, PS, gives much the same result. Fuel type is denoted by P for petrol, D for diesel. Engine configuration is a combination of the layout and the number of cylinders - S equates to straight (or in-line), F for flat (or horizontally opposed), V is self explanatory; the number of cylinders follows. The driven wheels are noted, Front, Rear or 4 wheel drive. Top speed and acceleration from rest to 60 mph are the two universally popular measures of a car's performance. Just one fuel consumption figure is given, for urban driving, as this is arguably the most realistic of the statutory tests. The insurance group is the standardised rating system used in the UK, on a scale of 1 to 20. Groups cannot be given for cars not sold in the UK, or the very latest models which have yet to be rated. Length and width are for the saloon or hatchback version, whichever is the longest. The weight, in kilograms, is for the lightest version in each range.

Finally, not all the cars listed will be available for sale in the UK. Check the new price tables later on for the definitive list of availability, then cross refer back to this list. Remember that some cars are sold only in a restricted model range in the UK, and different examples are available elsewhere - the data table is as complete as we can make it.

	cc	bhp	F	Engine configuration	D	mph	A	mpg	UK	Length	Width	Weight
ACURA												
CL												
2.2	2156	145	P	S4	F	119	-	25	-	4827	1780	1390
3.0	na	na	P	V6	F	-	-	-	-	4827	1780	1460
ALFA ROMEO												
145												
1.6	1596	103	P	F4	F	115	11.0	28	10	4095	1710	1140
1.7 16v	1712	129	P	F4	F	124	9.8	26	13	4095	1710	1190
2.0 16v	1970	150	P	S4	F	131	8.4	28	14	4095	1710	1240
146												
1.6	1596	103	P	F4	F	116	11.5	27	10	4255	1710	1175
1.7 16v	1712	129	P	F4	F	126	10.2	26	13	4255	1710	1225
2.0 16v	1970	150	P	S4	F	131	8.4	28	14	4255	1710	1275
155												
1.8 Twin Spark	1747	140	P	S4	F	127	10.0	30	14	4443	1730	1270
2.0 Twin Spark	1970	150	P	S4	F	129	9.3	29	15	4443	1730	1300
2.5	2492	163	P	V6	F	133	8.4	24	16	4443	1730	1370
164												
2.0 Twin Spark	1995	146	P	S4	F	130	9.9	29	13	4665	1760	1380
2.0 Turbo	1997	201	P	V6	F	148	8.2	24	-	4665	1760	1500
2.5 TD	2500	125	D	S4	F	126	10.8	33	-	4665	1760	1510
3.0 Super	2959	210	P	V6	F	149	8.0	21	16	4665	1760	1500
3.0 Cloverleaf	2959	230	P	V6	F	152	7.7	21	17	4665	1760	1680
Spider												
2.0 Twin Spark	1970	150	P	S4	F	131	8.4	27	17	4285	1780	1370
2.0 Turbo	1996	200	P	V6	F	147	7.2	24	-	4285	1780	1430
3.0	2959	192	P	V6	F	141	7.3	22	-	4285	1780	1420
GTV												
2.0 Twin Spark	1970	150	P	S4	F	134	8.4	27	16	4285	1780	1370
2.0 Turbo	1996	200	P	V6	F	147	7.2	24	-	4285	1780	1430
3.0	2959	192	P	V6	F	141	7.3	22	-	4285	1780	1420
ARO												
10												
1.4	1397	62	P	S4	4x4	75	-	23	-	3835	1645	1120
1.9 Diesel	1870	65	D	S4	4x4	81	-	26	-	3835	1645	1285
ASIA MOTORS												
Rocsta												
1.8	1789	77	P	S4	4x4	80	-	-	6	3580	1690	1310
2.2 Diesel	2184	61	D	S4	4x4	70	-	-	6	3580	1690	1360
ASTON MARTIN												
DB7	3239	335	P	S6	R	165	5.7	-	20	1830	1240	1700
V8	5430	350	P	V8	R	155	5.9	11	20	1945	1855	1920
Vantage	5430	557	P	V8	R	186	4.6	–	20	1945	1855	1990
AUDI												
A3												
1.6	1595	101	P	S4	F	117	11.0	29	-	4152	1735	1090
1.8	1781	125	P	S4	F	125	9.8	28	-	4152	1735	1120
1.8 T	1781	150	P	S4	F	135	7.9	29	-	4152	1735	1130
1.9 TDi	1896	110	D	S4	F	120	11.7	-	-	4152	1735	1180
A4												
1.6	1595	101	P	S4	F	119	11.9	29	10	4480	1735	1195
1.8	1781	125	P	S4	F	127	10.5	28	13	4480	1735	1225
1.8 T	1781	150	D	S4	F	140	8.3	29	15	4480	1735	1235
1.9 TDi	1896	90	D	S4	F	114	13.3	46	12	4480	1735	1240
1.9 TDi 100	1896	110	D	S4	F	122	11.3	-	13	4480	1735	1240
2.6	2598	150	P	V6	F	137	9.1	24	15	4480	1735	1285
2.6 quattro	2598	150	P	V6	4x4	135	9.1	22	15	4480	1735	1405
2.8	2771	174	P	V6	F	149	7.3	22	17	4480	1735	1285
2.8 quattro	2771	174	P	V6	4x4	148	7.3	22	17	4480	1735	1405
A6												
1.8	1781	125	P	S4	F	124	11.2	27	15	4795	1785	1370
1.9 TDi	1896	90	D	S4	F	110	13.9	40	14	4795	1785	1400
2.0	1984	115	P	S4	F	118	11.9	24	15	4795	1785	1345
2.5 TDi	2460	115	D	S5	F	121	11.1	38	15	4795	1785	1460
2.5 TDi 140bhp	2460	140	D	S5	F	129	9.9	38	16	4795	1785	1460
2.6	2598	150	P	V6	F	129	9.9	23	16	4795	1785	1440
2.8 quattro	2771	174	P	V6	4x4	135	9.1	20	17	4795	1785	1540
S6	2226	230	P	V6	4x4	150	6.7	19	19	4795	1785	1650
A8												
2.8	2771	174	P	V6	F	140	10.2	20	18	5035	1880	1460
3.7	3697	230	P	V8	F	153	8.7	19	19	5035	1880	1650
4.2 quattro	4172	300	P	V8	4x4	155	7.3	17	20	5035	1880	1750

KEY

CC **Engine** cc

Power bhp
F **Fuel**
Engine configuration
D **Driven wheels**
mph **Top speed** mph
A **0-60 mph** secs
mpg **Mpg urban**
UK **Insurance UK** group
Length mm
Width mm
Weight kg

Technical Data

	cc	bhp	F	●	D	mph	A	mpg	UK	⇕	⇔	▼
BENTLEY												
Brooklands	6750	300	P	V8	R	140	7.9	11	20	5268	1880	2430
Turbo R	6750	385	P	V8	R	150	5.9	11	20	5370	1880	2450
Continental R	6750	385	P	V8	R	155	5.9	11	20	5342	1880	2450
Azure	6750	385	P	V8	R	155	5.9	10	20	5342	1880	2450
BMW												
3-Series												
316i	1596	102	P	S4	R	117	12.3	31	10	4435	1710	1190
318i	1796	115	P	S4	R	125	11.3	29	11	4435	1710	1205
318i 16v	1895	140	P	S4	R	132	10.2	26	12	4435	1710	1240
320i	1991	150	P	S6	R	133	10.0	25	13	4435	1710	1285
323i	2494	170	P	S6	R	141	8.0	22	15	4435	1710	1310
328i	2793	193	P	S6	R	148	7.3	24	16	4435	1710	1320
M3	3201	321	P	S6	R	155	5.6	17	19	4435	1710	1440
318tds	1665	90	D	S4	R	114	12.0	37	10	4435	1710	1320
325td	2498	115	D	S6	R	123	12.0	32	12	4435	1710	1265
325tds	2498	143	D	S6	R	133	10.4	32	14	4435	1710	1320
5-Series												
520i	1991	150	P	S6	R	138	10.2	25	14	4775	1800	1410
523i	2495	170	P	S6	R	143	8.5	24	15	4775	1800	1420
528i	2793	193	P	S6	R	148	7.5	24	16	4775	1800	1440
535i	3498	235	P	V8	R	154	7.0	20	-	4775	1800	1540
540i	4398	286	P	V8	R	155	5.9	15	18	4775	1800	1585
525tds	2498	143	D	S6	R	132	10.4	32	14	4775	1800	1480
7-Series												
728i	2793	193	P	S6	R	140	9.6	18	17	4985	1860	1670
735i	3498	235	P	V8	R	151	8.4	19	18	4985	1860	1725
740i	4398	286	P	V8	R	155	7.6	17	19	4985	1860	1790
750i	5379	326	P	V12	R	155	6.6	14	20	4985	1860	1995
8-Series												
840Ci	4398	286	P	V8	R	155	7.4	16	20	4780	1855	1780
850CSi	5576	380	P	V12	R	155	6.0	14	20	4780	1855	1865
Z3												
1.8	1796	116	P	S4	R	121	10.5	30	-	4025	1690	1150
1.9	1895	140	P	S4	R	128	9.5	27	-	4025	1690	1175
BRISTOL												
Blenheim	5900	230	P	V8	R	140	6.9	-	20	4870	1750	1745
BUICK												
Century												
3.1	3135	162	P	V6	F	122	-	19	-	4185	1850	1520
Park Avenue												
3.8	3791	205	P	V6	F	125	-	18	-	5250	1895	1725
3.8 supercharged	3791	240	P	V6	F	138	-	16	-	5250	1895	1780
CADILLAC												
Catera												
3.0	2962	203	P	V6	R	138	8.5	20	-	4925	1785	1710
De Ville												
4.6	4565	275	P	V8	F	130	8.0	16	-	5325	1945	1708
4.6	4565	300	P	V8	F	130	7.5	16	-	5325	1945	1805
Seville												
4.6	4565	275	P	V8	F	131	7.5	16	-	5185	1885	1740
4.6	4565	300	P	V8	F	150	7.5	16	-	5185	1885	1765
CATERHAM												
Seven												
1.6	1588	117	P	S4	R	108	6.2	-	-	3380	1580	520
1.4 SS 6-speed	1588	140	P	S4	R	120	5.6	-	-	3380	1580	520
2.0 HPC	1998	165	P	S4	R	126	4.8	-	-	3380	1580	600
C21												
1.6	1588	117	P	S4	R	119	6.4	-	-	3800	1580	650
1.6 SS	1588	140	P	S4	R	131	5.8	-	-	3800	1580	650
CHEVROLET												
Camaro												
3.4	3791	200	P	V6	R	125	-	19	-	4910	1885	1500
5.7	5733	285	P	V8	T	150	6.5	14	-	4910	1885	1550

	cc	bhp	F	●	D	mph	A	mpg	UK	⇕	⇔	▼
CHEVROLET cont'd												
Corvette (96 model)												
5.7	5733	300	P	V8	R	159	4.9	15	-	4535	1795	1500
5.7	5735	330	P	V8	R	163	4.9	15	-	4535	1795	1500
Lumina												
3.1	3135	160	P	V6	F	125	10.0	19	-	4865	1880	1510
3.4 24v	3350	210	P	V6	F	131	8.4	19	-	4865	1880	1585
CHRYSLER												
Neon												
2.0	1996	132	P	S4	F	125	8.8	24	10	4365	1715	1100
2.0 DOHC	1996	150	P	S4	F	130	8.5	24	-	4365	1715	1100
Sebring												
2.0	1996	140	P	S4	F	125	10.9	24	-	4760	1770	1335
2.5	2497	155	P	V6	F	131	10.5	20	-	4760	1770	1420
Sebring Convertible												
2.4	2429	150	P	S4	F	-	-	23	-	4902	1780	1520
2.5	2497	168	P	V6	F	-	-	22	-	4902	1780	1540
Voyager												
2.4	2429	150	P	S4	F	120	-	20	-	5070	1950	1600
3.0	2972	150	P	V6	F	120	-	20	-	5070	1950	1708
3.3	3301	158	P	V6	F	120	-	18	-	5070	1950	1752
3.8	3778	166	P	V6	F	120	-	17	-	5070	1950	1792
2.5 TD	2500	115	D	S4	F	104	-	25	-	5070	1950	1850
Jeep Wrangler												
2.5	2464	122	P	S4	4x4	94	13.6	20	12	3860	1690	1335
4.0	3960	184	P	S6	4x4	100	8.8	16	14	3860	1690	1400
Jeep Cherokee												
2.5	2464	122	P	S4	4x4	103	12.1	20	13	4240	1790	1380
4.0	3960	184	P	S6	4x4	112	9.5	16	14	4240	1790	1440
2.5 TD	2499	115	D	S4	4x4	103	13.1	29	13	4240	1790	1470
Jeep Grand Cherokee												
4.0	3960	184	P	S6	4x4	113	10.2	15	16	4550	1800	1620
5.2	5210	220	P	V8	4x4	116	8.1	14	17	4550	1800	1795
CITROEN												
AX												
1.0	954	50	P	S4	F	93	13.5	44	3	3525	1555	690
1.1	1124	60	P	S4	F	104	10.6	40	5	3525	1555	690
1.4	1361	75	P	S4	F	108	9.3	35	9	3525	1555	770
1.5 Diesel	1527	58	D	S4	F	98	12.7	55	6	3525	1555	790
Saxo												
1.0	954	50	P	S4	F	93	19.1	-	-	3720	1595	805
1.1	1124	60	P	S4	F	102	14.5	32	4	3720	1595	805
1.4	1361	75	P	S4	F	109	11.9	30	5	3720	1595	840
1.6	1587	88	P	S4	F	116	11.6	-	-	3720	1595	905
1.6 16v	1587	118	P	S4	F	122	9.9	-	-	3720	1595	935
ZX												
1.1	1124	60	P	S4	F	100	16.8	36	-	4275	1685	935
1.4	1360	75	P	S4	F	107	11.0	33	6	4275	1685	945
1.8	1761	103	P	S4	F	117	9.5	30	11	4275	1685	1010
2.0	1998	123	P	S4	F	125	8.4	27	13	4275	1685	1060
2.0 16v	1998	150	P	S4	F	137	8.0	25	15	4275	1685	1150
1.9 Diesel	1905	71	D	S4	F	104	12.7	42	7	4275	1685	1025
1.9 TD	1905	92	D	S4	F	115	10.3	39	9	4275	1685	1085
Xantia												
1.6i	1580	89	P	S4	F	109	12.4	26	8	4445	1755	1170
1.8i 16v LX	1761	112	P	S4	F	120	10.0	27	10	4445	1755	1175
2.0i 16v LX	1998	135	P	S4	F	126	9.3	25	13	4445	1755	1240
2.0 Turbo	1998	147	P	S4	F	132	8.9	23	14	4445	1755	1375
1.9 Diesel	1905	71	D	S4	F	99	16.2	38	8	4445	1755	1210
1.9 TD	1905	92	D	S4	F	111	13.9	37	9	4445	1755	1250
2.1 TD	2088	109	D	S4	F	119	12.5	36	-	4445	1755	1385
XM												
2.0i 16v	1998	135	P	S4	F	127	9.1	24	14	4710	1795	1395
2.0i Turbo	1998	150	P	S4	F	133	8.2	22	15	4710	1795	1415
2.1 TD	2088	110	D	S4	F	119	11.0	36	14	4710	1795	1440
2.5 TD	2445	130	D	S4	F	124	10.4	31	14	4710	1795	1585
3.0 V6 auto	2963	170	P	V6	F	136	9.2	17	16	4710	1795	1495
Evasion/Synergy												
2.0	1998	123	P	S4	F	111	14.6	24	10	4455	1820	1510

Technical Data

	cc	bhp	F	●	D	mph	A	mpg	UK	⇕	⇔	▼
CITROEN cont'd												
2.0 Turbo	1998	150	P	S4	F	122	11.0	22	-	4455	1820	1575
1.9 TD	1905	92	D	S4	F	100	17.2	32	10	4455	1820	1565
DAEWOO												
Nexia												
1.5	1498	75	P	S4	F	101	12.3	29	4	4260	1660	915
1.5 16v	1498	90	P	S4	F	105	12.0	29	5	4260	1660	1005
Espero												
1.5	1498	90	P	S4	F	105	12.6	26	6	4615	1720	1085
1.8	1796	95	P	S4	F	112	10.8	24	7	4615	1720	1110
2.0	1998	105	P	S4	F	115	10.6	24	7	4615	1720	1195
DAIHATSU												
Charade												
1.3	1296	84	P	S4	F	105	11.2	37	8	4100	1620	820
1.5	1499	88	P	S4	F	105	10.8	37	10	4100	1620	840
1.6	1590	105	P	S4	F	115	9.6	38	11	4100	1620	875
Hijet												
1.0	993	47	P	S3	F	-	-	38	5	3295	1395	935
Fourtrak												
2.8 TD	2765	101	D	S4	4x4	84	-	30	8	4165	4165	1600
Sportrak												
1.6	1589	94	P	S4	4x4	93	-	27	10	3785	1635	1180
DODGE												
Intrepid												
3.3	3301	161	P	V6	F	120	11.0	18	-	5125	1890	1500
3.3	3518	214	P	V6	F	134	8.8	17	-	5125	1890	1530
Viper												
8.0	7990	450	P	V10	R	165	4.5	12	20	4450	1925	1588
DONKERVOORT												
D8												
1.8	1796	140	P	S4	R	125	6.0	-	-	3600	1730	570
1.8	1796	160	P	S4	R	131	5.0	-	-	3600	1730	570
2.0	1994	220	P	S4	R	147	4.5	-	-	3600	1730	660
2.0	1994	280	P	S4	R	153	4.0	-	-	3600	1730	695
FERRARI												
F355	3496	375	P	V8	R	175	4.5	12	20	4250	1900	1350
550M	5474	485	P	V12	R	199	4.3	-	20	4550	1935	1455
456	5474	436	P	V12	R	191	5.2	12	20	4730	1920	1790
F50	4698	513	P	V12	R	202	3.7	-	20	4480	1985	1230
FIAT												
Cinquecento												
900	899	41	P	S4	F	87	18.0	43	2	3225	1485	710
1.1	1108	54	P	S4	F	93	14.2	38	3	3225	1485	735
Panda												
750	770	34	P	S4	F	78	23	46	-	3410	1495	715
1.0	999	45	P	S4	F	88	16.0	43	-	3410	1495	715
Palio												
750	770	34	P	S4	F	78	23.0	46	1	3410	1495	715
1.0	999	45	P	S4	F	87	16.0	42	1	3410	1495	715
1.1	1108	50	P	S4	F	87	17.5	36	1	3410	1495	755
Punto												
1.1	1108	55	P	S4	F	93	16.5	36	3	3760	1625	840
1.2	1242	75	P	S4	F	106	12.0	36	5	3760	1625	875
1.4 GT	1372	136	P	S4	F	124	7.9	29	14	3760	1625	1000
1.6	1581	88	P	S4	F	110	11.5	30	7	3760	1625	965
1.7 Diesel	1698	57	D	S4	F	94	20.0	41	3	3760	1625	1000
1.7 TD	1698	72	D	S4	F	101	14.8	41	5	3760	1625	1010
Brava												
1.4	1370	80	P	S4	F	106	13.8	31	5	4025	1755	1040
1.6	1581	103	P	S4	F	112	11.5	30	7	4025	1755	1090
1.8	1747	113	P	S4	F	120	10.0	29	9	4025	1755	1130
2.0	1996	147	P	S5	F	131	8.5	26	-	4025	1755	1190
1.9 Diesel	1929	65	D	S4	F	97	17.8	43	-	4025	1755	1100
Bravo												
1.4	1370	80	P	S4	F	106	13.8	31	5	4185	1755	1040
FIAT cont'd												
1.6	1581	103	P	S4	F	112	11.5	30	7	4185	1755	1090
1.8	1747	113	P	S4	F	120	10.0	29	6	4185	1755	1130
1.9 Diesel	1929	65	D	S4	F	97	17.8	43	-	4185	1755	1130
Marea												
1.4	1370	80	P	S4	F	106	13.8	31	-	4380	1740	-
1.6	1581	103	P	S4	F	112	11.5	30	-	4380	1740	-
1.8	1747	113	P	S4	F	120	10.0	29	-	4380	1740	-
2.0	1996	147	P	S5	F	131	8.5	26	-	4380	1740	-
Barchetta												
1.8	1747	130	P	S4	F	125	8.9	29	-	3915	1640	1060
Coupe												
2.0 16v	1995	142	P	S4	F	129	9.2	25	17	4250	1765	1250
2.0 16v Turbo	1995	195	P	S4	F	140	7.5	23	18	4250	1765	1320
Ulysse												
2.0	1998	123	P	S4	F	111	14.6	24	10	4455	1820	1510
2.0 Turbo	1998	150	P	S4	F	122	11.0	22	-	4455	1820	1575
1.9 TD	1905	92	D	S4	F	100	17.2	32	10	4455	1820	1565
FORD												
Ka												
1.3	1299	60	P	S4	F	96	13.8	36	-	3620	1631	871
Fiesta												
1.3	1299	60	P	S4	F	96	14.8	37	4	3745	1605	949
1.25 16v	1242	75	P	S4	F	106	11.9	38	5	3745	1605	940
1.4 16v	1388	90	P	S4	F	112	10.8	34	7	3745	1605	940
1.8 Diesel	1753	60	D	S4	F	96	16.2	44	5	3745	1605	1022
Escort												
1.3	1299	60	P	S4	F	95	16.4	35	4	4290	1690	990
1.4	1392	71	P	S4	F	101	14.4	29	5	4290	1690	1030
1.6	1598	90	P	S4	F	110	11.5	31	6	4290	1690	1065
1.8	1753	105	P	S4	F	116	10.0	29	8	4290	1690	1065
2.0	1998	150	P	S4	F	130	8.0	25	16	4290	1690	1165
1.8 Diesel	1753	60	D	S4	F	95	16.7	44	6	4290	1690	1080
1.8 TD	1753	90	D	S4	F	106	10.8	38	7	4290	1690	1115
Mondeo												
1.6i	1598	90	P	S4	F	112	12.9	28	7	4556	1745	1295
1.8i	1796	115	P	S4	F	121	10.2	25	9	4556	1745	1295
2.0i	1989	130	P	S4	F	128	9.4	25	11	4556	1745	1315
2.5i	2544	170	P	V6	F	139	8.0	21	15	4556	1745	1390
1.8 TD	1753	90	D	S4	F	112	12.3	32	8	4556	1745	1360
Scorpio												
2.0 8v	1998	113	P	S4	F	121	12.8	25	12	4825	1760	1435
2.0 16v	1998	134	P	S4	F	129	11.1	24	13	4825	1760	1490
2.3 16v	2295	147	P	S4	F	130	9.2	24	15	4825	1760	1441
2.9 24v	2935	207	P	V6	F	141	9.0	20	15	4825	1760	1545
2.5 TD	2500	113	D	S4	F	121	11.9	27	15	4825	1760	1545
Probe												
2.0	1991	115	P	S4	F	127	10.6	30	16	4560	1775	1220
2.5	2497	165	P	V6	F	136	8.5	23	17	4560	1775	1325
Galaxy												
2.0	1998	115	P	S4	F	110	12.0	24	11	4615	1810	1560
2.8	2792	174	P	V6	F	124	10.0	21	14	4615	1810	1670
1.9 TD	1896	90	D	S4	F	100	15.4	37	11	4615	1810	1670
Maverick												
2.4	2389	124	P	S4	4x4	99	14.0	21	10	4105	1735	1620
2.7 TD	2663	125	D	S4	4x4	99	15.8	28	10	4105	1735	1730
FORD US												
Escort												
2.0	1988	110	P	S4	F	115	-	28	-	4435	4385	1160
Contour												
2.0	1988	125	P	S4	F	122	10.6	22	-	4670	1755	1256
2.5	2544	170	P	V6	F	140	8.6	20	-	4670	1755	1355
Taurus												
3.0	2986	145	P	V6	F	112	-	20	-	5015	1855	1510
3.0-24v	2967	200	P	V6	F	131	-	19	-	5015	1855	1515
Mustang												
3.8	3797	150	P	V6	R	112	9.3	20	-	4610	1825	1390
4.6	4601	215	P	V8	R	134	-	17	-	4610	1825	1390
4.6 DOHC	4601	305	P	V8	R	150	5.5	18	-	4610	1825	1390

	cc	bhp	F	●	D	mph	A	mpg	UK	⬍	⬌	▼
FORD US cont'd												
Expedition												
4.6	4601	215	P	V8	4x4	-	-	14	-	5195	1995	-
5.4	-	230	P	V8	4x4	-	-	13	-	5195	1995	-
FORD Australia												
Falcon												
4.0	3984	213	P	S6	R	131	8.0	24	-	4900	1855	1500
5.0	4942	224	P	V8	R	138	-	19	-	4900	1855	1600
5.0	4942	272	P	V8	R	144	7.1	14	-	4900	1855	1645
FSO												
Polonez/Caro												
1.5	1481	72	P	S4	R	93	18.8	28	5	4320	1650	1115
1.9 Diesel	1905	68	D	S4	R	87	21.8	40	6	4320	1650	1135
HINDUSTAN												
Ambassador												
1.5	1488	54	P	S4	R	81	-	-	-	4310	1675	1165
1.8	1817	73	P	S4	R	88	-	-	-	4310	1675	1165
Contessa												
1.8	1817	88	P	S4	R	99	-	-	-	4590	1700	1140
HOLDEN												
Commodore												
3.8	3791	200	P	V6	R	125	8.7	19	-	4860	1795	1360
5.0	4987	225	P	V8	R	131	8.3	17	-	4860	1795	1420
5.0 SS	4987	252	P	V8	R	144	8.0	17	-	4860	1795	1420
5.7 HSV	5733	292	P	V8	R	156	6.5	15	-	4860	1795	1600
HONDA												
Civic												
1.4	1396	75	P	S4	F	103	13.9	38	-	4395	1695	1030
1.4	1396	89	P	S4	F	111	13.1	34	9	4395	1695	1075
1.5	1493	89	P	S4	F	112	13.0	40	9	4395	1695	1080
1.6	1590	111	P	S4	F	119	10.2	32	11	4395	1695	1125
1.6 SR	1590	124	P	S4	F	122	9.9	32	13	4395	1695	1125
1.6 VTi	1595	160	P	S4	F	129	8.0	30	15	4395	1695	1165
CRX												
1.6	1590	123	P	S4	F	118	9.3	32	14	4000	1695	1035
CRV												
2.0	1973	131	P	S4	4X4	100	-	-	-	4470	1750	1340
Accord												
1.8	1850	115	P	S4	F	121	11.3	27	10	4685	1715	1275
2.0	1997	131	P	S4	F	124	10.1	27	11	4685	1715	1325
2.2	2156	150	P	S4	F	130	9.0	27	13	4685	1715	1370
2.7	2675	173	P	V6	F	131	-	24	-	4685	1715	1460
2.0 D	1994	105	P	S4	F	115	11.6	49	10	4685	1715	1345
Shuttle												
2.2 auto	2156	150	P	S4	F	114	12.2	23	16	4750	1770	1470
Prelude												
2.0	1997	134	P	S4	F	126	9.2	27	-	4545	1750	1240
2.2	2157	185	P	S4	F	143	7.5	26	-	4545	1750	1319
Legend												
3.5	3474	205	P	V6	F	134	9.1	17	17	4980	1810	1675
NSX												
3.0	2977	270	P	V6	R	160	5.9	22	20	4430	1810	1370
HUMMER												
Estate 5.7	5733	190	P	V8	4x4	83	19.5	-	-	4686	2197	2989
Estate 6.5 D	6466	170	D	V8	4x4	83	19.5	-	-	4686	2197	3039
Estate 6.5 TD	6466	190	D	V8	4x4	83	18.0	-	-	4686	2197	3070
HYUNDAI												
Accent												
1.3	1341	84	P	S4	F	108	12.6	39	6	4115	1620	935
1.5	1495	88	P	S4	F	109	11.4	36	8	4115	1620	965
Lantra												
1.6	1599	112	P	S4	F	120	10.9	26	8	4420	1700	1130
1.8	1795	126	P	S4	F	122	9.1	26	11	4420	1700	1200

	cc	bhp	F	●	D	mph	A	mpg	UK	⬍	⬌	▼
HYUNDAI cont'd												
Coupe												
2.0	1975	137	P	S4	F	123	10.7	25	-	4340	1730	-
Sonata												
1.8	1796	98	P	S4	F	103	12.1	28	-	4700	1770	1275
2.0	1997	105	P	S4	F	106	11.9	32	-	4700	1770	1275
2.0	1997	136	P	S4	F	121	10.2	33	13	4700	1770	1300
3.0	2972	143	P	V6	F	115	10.1	24	14	4700	1770	1400
Grandeur												
2.0	1997	145	P	S4	F	116	-	-	-	4980	1810	1520
3.0	2972	205	P	V6	F	118	-	-	-	4980	1810	1660
ISUZU												
Trooper												
3.2	3165	174	P	V6	4x4	106	11.5	16	13	4270	1745	1795
3.1 TD	3059	113	D	S4	4x4	94	16.6	24	13	4270	1745	1850
JAGUAR												
XJ												
XJ6 3.2	3239	219	P	S6	R	138	7.9	19	16	5025	1800	1800
XJ6 4.0	3980	249	P	S6	R	143	7.0	17	18	5025	1800	1800
XJR 4.0	3980	326	P	S6	R	155	5.9	17	20	5025	1800	1875
XJ12 6.0	5993	318	P	V12	R	155	6.8	14	20	5025	1800	1975
XK8												
4.0	3996	290	P	V8	R	155	-	-	-	-	-	-
KIA												
Pride												
1.1	1139	56	P	S4	F	94	13.6	37	-	3475	1605	700
1.3	1324	64	P	S4	F	100	11.8	35	5	3475	1605	750
Mentor												
1.6	1598	79	P	S4	F	107	11.8	29	8	4360	1690	1030
1.8	1840	124	P	S4	F	119	-	-	-	4360	1690	1030
Avella												
1.3	1323	76	P	S4	F	100	12.5	-	-	3825	1670	880
1.5	1498	100	P	S4	F	112	11.5	-	-	3825	1670	915
Sportage												
2.0	1998	95	P	S4	4x4	97	18.4	21	-	3760	1730	1420
2.0	1998	126	P	S4	4x4	103	14.7	22	9	3760	1730	1420
2.2 D	2184	62	D	S4	4x4	81	-	-	-	3760	1730	1465
2.0 TD	1998	83	D	S4	4x4	89	-	-	-	3760	1730	1465
LADA												
Samara												
1.1	1099	53	P	S4	F	85	14.1	39	5	4005	1620	900
1.3	1288	65	P	S4	F	92	14.5	30	6	4005	1620	915
1.5	1499	75	P	S4	F	100	12.0	30	6	4005	1620	930
Riva												
1.2	1198	60	P	S4	R	88	17.0	31	-	4045	1610	955
1.3	1294	65	P	S4	R	91	15.9	30	-	4045	1610	955
1.5	1452	67	P	S4	R	95	14.0	29	6	4045	1610	995
Niva												
1.6	1568	78	P	S4	4x4	82	22.0	24	8	3720	1676	1150
1.9 D	1905	64	D	S4	4x4	78	25.0	33	-	3720	1676	1180
LAMBORGHINI												
Diablo												
5.7	5707	492	P	V12	R	202	4.1	14	20	4460	2040	1575
5.7 SV	5707	500	P	V12	R	186	3.9	14	20	4470	2040	1530
LANCIA												
Y10												
1.2	1242	60	P	S4	F	100	13.3	37	-	3735	1690	860
1.4	1371	80	P	S4	F	106	12.4	32	-	3735	1690	920
Delta												
1.6	1581	75	P	S4	F	108	13.8	27	-	4010	1705	1120
1.8	1756	103	P	S4	F	116	11.8	25	-	4010	1705	1200
2.0	1995	139	P	S4	F	129	9.6	25	-	4010	1705	1250
2.0 Turbo	1995	186	P	S4	F	138	7.5	24	-	4010	1705	1330
1.9 TD	1929	90	D	S4	F	113	12.0	40	-	4010	1705	1280

	cc	bhp	F	●	D	mph	A	mpg	UK	⬍	◀▶	▼
LANCIA cont'd												
Dedra												
1.6	1581	90	P	S4	F	113	13.4	29	-	4345	1700	1140
1.8	1756	101	P	S4	F	116	12.9	28	-	4345	1700	1215
2.0	1995	113	P	S4	F	116	12.5	23	-	4345	1700	1250
2.0 16v	1995	139	P	S4	F	131	9.4	25	-	4345	1700	1260
1.9 TD	1929	90	D	S4	F	113	12.9	40	-	4345	1700	1225
Kappa												
2.0	1998	145	P	S5	F	128	9.8	24	-	4685	1825	1440
2.0 Turbo	1995	205	P	S4	F	147	7.3	25	-	4685	1825	1480
2.4	2446	175	P	S5	F	134	9.2	21	-	4685	1825	1450
3.0	2959	204	P	V6	F	141	8.0	20	-	4685	1825	1510
2.4 TD	2387	124	D	S5	F	121	11.5	30	-	4685	1825	1485
Z												
2.0	1998	123	P	S4	F	111	14.6	24	10	4455	1820	1510
2.0 Turbo	1998	150	P	S4	F	122	11.0	22	-	4455	1820	1575
1.9 TD	1905	92	D	S4	F	100	17.2	32	10	4455	1820	1565
LAND ROVER												
Defender												
2.5 TDi 90	2495	111	D	S4	4x4	85	15.1	28	7	3720	1790	1695
2.5 TDi 110	2495	111	D	S4	4x4	80	17.4	29	7	3720	1790	1870
Discovery												
2.0 Mpi	1994	134	P	S4	4x4	98	15.3	20	11	4520	1810	1890
2.5 Tdi	2495	111	D	S4	4x4	91	17.2	33	12	4520	1810	2055
3.9 V8i	3947	182	P	V8	4x4	106	10.8	14	13	4520	1810	1920
Range Rover												
2.5 TD	2497	134	D	S6	4x4	105	13.3	26	14	4715	1850	2115
4.0 V8	3947	190	P	V8	4x4	118	9.9	15	14	4715	1850	2090
4.6 V8	4554	225	P	V8	4x4	125	9.3	13	16	4715	1850	2150
LEXUS												
GS300	2997	209	P	S6	R	143	8.6	20	17	4965	1795	1720
LS400	3969	260	P	V8	R	156	7.4	17	19	4995	1830	1680
LINCOLN												
Continental												
4.6	4601	260	P	V8	F	134	-	15	-	5240	1870	1765
Mark VIII												
4.6	4601	280	P	V8	R	134	-	14	-	5265	1900	1710
LISTER												
Storm	6996	594	P	V12	R	200	4.1	-	20	4547	1975	1438
LOTUS												
Elise												
1.8	1795	116	P	S4	R	126	5.9	29	18	3726	1701	690
Esprit												
S4	2174	264	P	S4	R	160	4.7	18	19	4370	1865	1340
S4S	2174	300	P	S4	R	165	4.6	19	20	4370	1865	1340
V8	3506	349	P	V8	R	176	4.5	-	20	4414	1883	1380
MAHINDRA												
CJ3	2112	60	D	S4	4x4	65	-	-	7	4060	1650	1200
MARCOS												
Mantara												
Mantara 4.0	3950	190	P	V8	R	140	5.4	-	20	4005	1680	1080
Mantara 5.0	4998	320	P	V8	R	165	4.6	-	20	4005	1680	1080
LM 4.0	3950	190	P	V8	R	143	5.4	-	20	4263	1828	1120
LM 5.0	4998	320	P	V8	R	168	4.6	-	20	4263	1828	1120
MASERATI												
Ghibli												
2.0	1996	306	P	V6	R	166	5.7	15	20	4225	1775	1365
2.0 Cup	1996	330	P	V6	R	169	5.6	-	-	4225	1775	1300
2.8	2790	284	P	V6	R	163	5.7	16	20	4550	1810	1365
Quattroporte												
2.0	1996	287	P	V6	R	163	5.9	18	20	4550	1810	1545
2.8	2790	284	P	V6	R	159	5.9	15	20	4550	1810	1560
3.2	3217	335	P	V8	R	169	5.8	14	20	4550	1810	1650

	cc	bhp	F	●	D	mph	A	mpg	UK	⬍	◀▶	▼
MAZDA												
121												
1.3	1299	60	P	S4	F	96	14.8	37	4	3745	1605	949
1.25 16v	1242	75	P	S4	F	106	11.9	38	5	3745	1605	940
1.8 Diesel	1753	60	D	S4	F	96	16.2	44	5	3745	1605	1022
323												
1.3i	1324	75	P	S4	F	101	13.3	35	7	4340	1695	1010
1.5i	1498	90	P	S4	F	108	11.9	30	9	4340	1695	1080
1.8i	1840	115	P	S4	F	119	10.0	29	12	4340	1695	1110
2.0i	1995	147	P	V6	F	129	9.4	25	14	4340	1695	1200
626												
1.8i	1840	105	P	S4	F	119	11.9	31	14	4695	1750	1140
1.8i	1845	140	P	V6	F	125	-	-	-	4695	1750	1140
2.0i	1991	115	P	S4	F	124	10.4	29	14	4695	1750	1140
2.5i	2497	165	P	V6	F	138	8.5	24	-	4695	1750	1140
2.0 Diesel	1998	75	D	S4	F	101	14.7	35	14	4695	1750	1255
Xedos 6												
1.6i	1598	114	P	S4	F	110	10.3	28	14	4560	1695	1160
2.0i	1995	146	P	V6	F	134	9.3	26	16	4560	1695	1200
Xedos 9												
2.5	2497	168	P	V6	F	130	11.0	22	15	4825	1770	1415
MX-3												
1.6i	1598	108	P	S4	F	105	14.0	29	12	4220	1695	1060
1.8i	1845	134	P	V6	F	126	8.5	25	15	4220	1695	1115
MX-5												
1.6i	1598	88	P	S4	R	109	10.6	30	11	3950	1675	990
1.8i	1839	134	P	S4	R	123	8.7	28	12	3950	1675	990
MX-6												
2.5i	2497	165	P	V6	F	136	8.5	21	16	4625	1750	1170
MCLAREN												
F1	6064	627	P	V12	R	235	3.2	-	20	4290	1820	1140
MERCEDES-BENZ												
C-Class												
C180	1799	122	P	S4	R	120	12.1	26	11	4485	1720	1280
C200	1998	136	P	S4	R	123	11.1	25	12	4485	1720	1295
C230	2295	150	P	S4	R	131	10.6	24	13	4485	1720	1340
C230 K	2250	193	P	S4	R	144	8.4	27	13	4485	1720	1340
C280	2799	193	P	S6	R	143	8.5	20	15	4485	1720	1420
C36 AMG	3606	280	P	S6	R	155	6.9	21	-	4485	1720	1490
C220 D	2155	95	D	S4	R	109	16.3	33	11	4485	1720	1310
C250 TD	2497	150	D	S4	R	127	10.2	30	13	4485	1720	1410
E-Class												
E200	1998	136	P	S4	R	128	11.3	26	13	4800	1800	1440
E230	2295	150	P	S4	R	134	10.5	25	14	4800	1800	1450
E280	2799	193	P	S6	R	144	9.1	19	15	4800	1800	1570
E320	3199	220	P	S6	R	147	7.8	22	16	4800	1800	1600
E420	4196	279	P	V8	R	156	7.0	20	-	4800	1800	1690
E50 AMG	4973	347	P	V8	R	156	6.2	18	-	4800	1800	1680
E220 D	2155	95	D	S4	R	113	17.0	34	13	4800	1800	1460
E290 TD	2874	129	D	S4	R	122	11.5	36	-	4800	1800	1540
E300 D	2996	136	D	S4	R	128	13.0	28	14	4800	1800	1560
E220 Cabrio	2199	150	P	S4	R	131	10.6	24	17	4800	1800	1620
E320 Cabrio	3199	220	P	S6	R	141	7.9	18	18	4800	1800	1710
S-Class												
S280	2799	193	P	S6	R	130	10.7	19	16	5215	1885	1890
S320	3199	231	P	S6	R	131	8.9	20	17	5215	1885	1890
S420	4196	268	P	V8	R	152	8.3	18	18	5215	1885	1990
S500	4973	308	P	V8	R	155	7.3	16	19	5215	1885	2000
S600L	5987	389	P	V12	R	155	6.3	14	20	5215	1885	2190
S500 Coupe	4973	308	P	V8	R	155	7.3	16	20	5065	1910	2080
S600 Coupe	5987	389	P	V12	R	155	6.6	14	20	5065	1910	2240
SLK												
2.0	1998	136	P	S4	R	129	9.5	-	-	3995	1715	1260
2.3K	2295	103	P	S4	R	142	7.4	-	-	3995	1715	1325
SL												
SL280	2799	193	P	S6	R	140	9.5	21	20	4470	1810	1760
SL320	3199	231	P	S6	R	149	8.1	20	20	4470	1810	1780
SL500	4973	320	P	V8	R	155	6.2	18	20	4470	1810	1800
SL600	5987	389	P	V12	R	155	6.1	14	20	4470	1810	1980

Technical Data

MERCURY

	cc	bhp	F	●	D	mph	A	mpg	UK	⬍	⬌	▼
Mountainer												
5.0	4942	211	P	V8	4x4	112	-	14	-	4790	1785	1750
MGF												
1.8i	1796	120	P	S4	R	120	8.5	34	13	3915	1630	1055
1.8i VVC	1796	145	P	S4	R	130	7.0	30	16	3915	1630	1065

MITSUBISHI

	cc	bhp	F	●	D	mph	A	mpg	UK	⬍	⬌	▼
Colt												
1.3	1299	75	P	S4	F	106	12.5	36	5	4290	1690	940
1.6	1597	90	P	S4	F	116	10.5	33	6	4290	1690	97 5
Carisma												
1.6	1597	90	P	S4	F	113	12.0	33	33	4435	1695	1105
1.8	1834	115	P	S4	F	125	10.2	32	33	4435	1695	1130
1.8 DOHC	1834	140	P	S4	F	134	9.2	31	35	4435	1695	1175
Galant												
Data on new model unavailable as we went to press												
Space Runner												
1.8	1834	121	P	S4	F	113	10.5	28	13	4290	1695	1185
Space Wagon												
2.0	1997	131	P	S4	F	116	11.2	26	13	4515	1695	1300
2.0 TD	1998	80	P	S4	F	97	17.8	31	12	4515	1695	1330
Space Gear												
2.0	1997	115	P	S4	R	103	-	20	-	5085	1695	1560
2.4	2351	132	P	S4	R	103	-	20	-	5085	1695	1660
3.0	2972	185	P	V6	R	106	-	18	-	5085	1695	2030
2.5 TD	2477	99	D	S4	R	93	-	20	-	5085	1695	1650
2.8 TD	2835	125	D	S4	R	94	-	19	-	5085	1695	2020
Pajero Junior												
700	659	54	P	S4	4x4	72	-	40	-	3295	1395	850
700 20v	659	64	P	S4	4x4	81	-	31	-	3295	1395	890
1100	1095	80	P	S4	4x4	84	-	31	-	3295	1395	960
Shogun												
2.5 TD	2477	98	D	S4	4x4	91	18.1	26	13	4656	1695	1550
2.8 TD	2835	123	D	S4	4x4	94	17.3	21	13	4656	1695	1850
3.0	2972	178	P	V6	4x4	109	11.1	18	14	4656	1695	1740
3.5	3497	205	P	V6	4x4	116	10.5	16	15	4656	1695	1890
FTO												
1.8	1834	125	P	S4	F	125	-	26	-	4320	1735	1100
2.0	1998	170	P	V6	F	138	-	22	-	4320	1735	1150
2.0 s'ch'rg'd	1998	200	P	V6	F	144	-	22	-	4320	1735	1150
3000												
GT	2972	281	P	V6	4x4	155	5.9	20	20	4570	1840	1710

MORGAN

	cc	bhp	F	●	D	mph	A	mpg	UK	⬍	⬌	▼
4/4												
1.8	1796	121	P	S4	R	-	-	n/a	13	3890	1500	870
Plus 4												
2.0	1994	134	P	S4	R	115	8.1	n/a	14	3960	1630	920
Plus 8												
3.9	3946	185	P	V6	R	130	5	n/a	15	3999	1700	940

MOSKVICH

	cc	bhp	F	●	D	mph	A	mpg	UK	⬍	⬌	▼
2141												
1.6	1569	76	P	S4	R	97	15.7	28	-	4350	1690	1080
1.7	1702	85	P	S4	R	100	14.9	28	-	4350	1690	1080
1.8 Diesel	1753	60	D	S4	R	87	22.0	43	-	4350	1690	1140

NISSAN

	cc	bhp	F	●	D	mph	A	mpg	UK	⬍	⬌	▼
Micra												
1.0	998	54	P	S4	F	93	16.4	47	3	3700	1590	775
1.3	1275	75	P	S4	F	106	12.0	43	5	3700	1590	810
Almera												
1.4	1392	87	P	S4	F	108	12.6	35	5	4440	1690	1035
1.6	1597	100	P	S4	F	112	11.0	33	6	4440	1690	1070
2.0	1998	143	P	S4	F	131	8.2	26	14	4120	1690	1070
2.0 Diesel	1974	75	D	S4	F	98	16.8	41	5	4440	1690	1140
Primera												
1.6	1597	100	P	S4	F	112	12.0	30	-	4430	1715	1165
2.0	1998	130	P	S4	F	127	9.6	26	-	4430	1715	1205

NISSAN cont'd

	cc	bhp	F	●	D	mph	A	mpg	UK	⬍	⬌	▼
2.0 TD	1974	90	D	S4	F	108	14.0	33	-	4430	1715	1260
QX												
2.0	1995	138	P	V6	F	126	11.3	26	12	4770	1770	1310
3.0	2988	190	P	V6	F	131	9.6	22	14	4770	1770	1360
Infiniti												
4.1	4130	266	P	V8	R	144	-	18	-	5060	1820	1765
200 SX												
2.0	1998	160	P	S4	R	131	-	-	-	4520	1730	1140
2.0 Turbo	1998	203	P	S4	R	146	7.5	27	17	4520	1730	1260
Skyline												
2.0	1998	130	P	S6	R	125	-	20	-	4720	1720	1270
2.5	2498	200	P	S6	R	131	-	19	-	4720	1720	1360
2.6 Turbo	2569	280	P	S6	4x4	150	-	17	-	4720	1720	1540
Rasheen												
1.5	1498	105	P	S4	F	94	-	24	-	3980	1695	1160
Serena												
1.6	1597	97	P	S4	F	93	18	25	10	4315	1695	1385
2.0	1998	126	P	S4	F	106	13	22	12	4315	1695	1485
2.3 Diesel	2283	75	D	S4	F	84	27	29	10	4315	1695	1485
Terrano II												
2.4	2389	124	P	S4	4x4	99	13.7	19	10	4585	1735	1620
2.7 TD	2663	125	D	S4	4x4	97	16.7	23	10	4585	1735	1730
Patrol												
4.2	4169	170	P	S6	4x4	106	14.3	14	12	4845	1930	2170
2.8 TD	2826	116	D	S6	4x4	91	19.3	23	12	4845	1930	2180

OLDSMOBILE

	cc	bhp	F	●	D	mph	A	mpg	UK	⬍	⬌	▼
Cutlass												
3.1	3135	160	P	V6	F	-	-	-	-	4877	1763	1355
Bravada												
4.3	4300	190	P	V6	4x4	112	10.0	16	-	4595	1690	1825

PEUGEOT

	cc	bhp	F	●	D	mph	A	mpg	UK	⬍	⬌	▼
106												
1.0	954	50	P	S4	F	93	19.1	-	-	3720	1595	805
1.1	1124	60	P	S4	F	102	14.5	32	4	3720	1595	805
1.4	1361	75	P	S4	F	109	11.9	30	5	3720	1595	840
1.6	1587	88	P	S4	F	116	11.6	-	-	3720	1595	905
1.6 16v	1587	118	P	S4	F	122	9.9	-	-	3720	1595	935
306												
1.4	1360	75	P	S4	F	102	14.9	34	4	4230	1690	1020
1.6	1597	90	P	S4	F	111	12.9	31	5	4230	1690	1060
1.8	1762	102	P	S4	F	114	12.2	27	5	4230	1690	1080
2.0	1998	123	P	S4	F	122	9.2	34	12	4230	1690	1140
2.0 GTi-6	1998	155	P	S4	F	133	8.4	24	15	4230	1690	1160
1.9 Diesel	1905	70	D	S4	F	101	16.9	40	4	4230	1690	1080
1.9 TD	1905	92	D	S4	F	112	12.4	37	5	4230	1690	1120
406												
1.6	1580	90	P	S4	F	109	15.3	28	-	4560	1765	1240
1.8	1762	112	P	S4	F	120	12.5	28	10	4560	1765	1275
2.0	1998	135	P	S4	F	127	11.0	25	12	4560	1765	1315
2.0 Turbo	1998	150	P	S4	F	130	10.3	20	13	4560	1765	1315
1.9 TD	1905	92	D	S4	F	111	14.3	35	9	4560	1765	1335
2.1 TD	2000	110	D	S4	F	119	12.5	35	11	4560	1765	1415
605												
2.0 base	1998	122	P	S4	F	127	10.9	24	14	4765	1800	1415
2.0	1998	145	P	S4	F	132	10.0	22	15	4765	1800	1485
2.1 Diesel	2088	110	D	S4	F	119	13.1	36	14	4765	1800	1360
3.0	2963	110	P	V6	F	134	10.9	17	16	4765	1800	1570
3.0 24v	2963	200	P	V6	F	147	8.6	18	17	4765	1800	1650
806												
2.0	1998	123	P	S4	F	111	14.6	24	10	4455	1820	1510
2.0 Turbo	1998	150	P	S4	F	122	11.0	22	-	4455	1820	1575
1.9 TD	1905	92	D	S4	F	100	17.2	32	10	4455	1820	1565

PLYMOUTH

	cc	bhp	F	●	D	mph	A	mpg	UK	⬍	⬌	▼
Prowler												
3.5	3518	214	P	V6	R	-	-	-	-	4190	1930	1350
Breeze												
2.0	1996	132	P	S4	F	119	10.0	24	-	4730	1820	1330

	cc	bhp	F	●	D	mph	A	mpg	UK	⇔	⇕	▼
PONTIAC												
Grand Prix												
3.1	3135	160	P	V6	F	122	-	18	-	4990	1845	1540
3.8	3791	195	P	V6	F	125	-	16	-	4990	1845	1540
3.8 supercharged	3791	245	P	V6	F	141	-	16	-	4990	1845	1540
Sunfire												
2.2	2190	122	P	S4	F	109	-	24	-	4615	1725	1210
2.4	2392	150	P	S4	F	119	-	22	-	4615	1725	1270
Bonneville												
3.8	3791	205	P	V6	F	125	-	18	-	5125	1890	1575
3.8 supercharged	3791	240	P	V6	F	144	8.5	16	-	5125	1890	1595
PORSCHE												
911												
Carrera	3600	272	P	F6	R	168	5.6	16	20	4245	1735	1370
Turbo	3600	408	P	F6	4x4	180	4.5	14	20	4245	1795	1500
PREMIER												
118E												
1.2	1171	52	P	S4	R	85	-	29	-	4050	1610	900
1.4 Diesel	1366	41	D	S4	R	75	-	35	-	4050	1610	965
PROTON												
Compact												
1.3	1299	75	P	S4	F	103	13.6	29	8	3990	1700	930
1.5	1468	89	P	S4	F	108	12.1	28	9	3990	1700	940
1.6	1597	111	P	S4	F	116	10.8	26	11	3990	1700	160
Persona												
1.5	1468	89	P	S4	F	108	12.1	33	9	4360	1690	975
1.6	1597	111	P	S4	F	116	10.8	30	11	4360	1690	1045
1.8	1834	116	P	S4	F	119	10.5	24	13	4360	1690	1065
1.8 16v	1834	133	P	S4	F	126	9.0	25	-	4360	1690	1100
2.0D	1998	68	D	S4	F	99	18.5	30	11	4360	1690	1115
RENAULT												
Twingo												
1.2	1239	55	P	S4	F	94	14.0	38	-	3435	1630	790
Clio												
1.2	1149	60	P	S4	F	100	13.5	41	3	3710	1630	825
1.4	1390	80	P	S4	F	109	11.0	36	5	3710	1630	850
1.8	1794	110	P	S4	F	122	9.0	27	11	3710	1630	925
1.9 Diesel	1870	65	D	S4	F	100	14.5	43	4	3710	1630	910
Megane												
1.4	1390	70	P	S4	F	105	14.3	41	3	4130	1700	1015
1.6	1598	75	P	S4	F	109	-	35	-	4130	1700	1055
1.6e	1598	90	P	S4	F	116	11.5	33	5	4130	1700	1055
2.0	1998	113	P	S4	F	125	9.7	27	8	4130	1700	1085
2.0 16v	1998	150	P	S4	F	134	8.6	27	13	3930	1700	1095
1.9 Diesel	1870	64	D	S4	F	100	16.5	39	4	4130	1700	1110
1.9 TD	1870	90	D	S4	F	113	12.3	40	6	4130	1700	1130
Laguna												
1.8	1794	95	P	S4	F	112	14.0	26	7	4620	1750	1225
2.0	1998	115	P	S4	F	124	10.6	25	10	4620	1750	1245
2.0 16v	1998	140	P	S4	F	127	9.8	26	13	4620	1750	1400
2.2 Diesel	2188	85	D	S4	F	109	15.3	35	9	4620	1750	1335
3.0 V6	2963	170	P	S4	F	137	9.2	17	15	4620	1750	1370
Safrane												
2.0	1948	138	P	S4	F	-	-	-	-	4750	1820	1370
2.5	2435	168	P	S5	F	-	-	-	-	4750	1820	1410
3.0 V6i	2975	170	P	V6	F	132	10.2	17	15	4750	1820	1465
2.2 TD	2188	115	D	S4	F	-	-	-	-	4750	1820	1410
Espace (1996 model)												
2.0	1995	105	P	S4	F	107	12.2	28	14	4430	1790	1330
2.2 auto	2165	110	P	S4	F	106	12.8	28	14	4430	1790	1330
2.9 auto	2849	153	P	S4	F	118	10.6	18	15	4430	1790	1420
2.1 TD	2068	88	D	S4	F	104	14.5	34	14	4430	1790	1340
Sport Spider												
2.0	1998	150	P	S4	R	131	6.5	-	-	3795	1830	790

	cc	bhp	F	●	D	mph	A	mpg	UK	⇔	⇕	▼
ROLLS-ROYCE												
Silver Dawn	6750	245	P	V8	R	133	9.5	11	20	5395	2110	2470
Silver Spur	6750	300	P	V8	R	140	7.9	11	20	5395	2110	2470
ROVER												
Mini												
1.3 Cooper	1275	63	P	S4	F	92	11.5	37	6	3050	1410	695
100												
111	1120	60	P	S4	F	96	13.7	40	3	3570	1560	815
114	1396	74	P	S4	F	103	10.7	35	5	3570	1560	840
115 Diesel	1527	56	D	S4	F	96	15.3	50	4	3570	1560	840
200												
214	1396	75	P	S4	F	103	12.5	36	4	3970	1690	985
214 16v	1396	103	P	S4	F	115	10.2	35	6	3970	1690	1015
216	1590	111	P	S4	F	118	9.3	33	7	3970	1690	1025
200 vi	1795	145	P	S4	F	127	7.5	30	14	3970	1690	1060
220D	1994	105	D	S4	F	105	12.0	47	5	3970	1690	1105
Cabriolet												
216	1590	111	P	S4	F	115	9.7	31	11	4220	1680	1135
Tourer												
216	1590	111	P	S4	F	115	9.7	31	9	4370	1680	1120
200 vi	1795	145	P	S4	F	125	8.0	28	12	4370	1680	1080
218 TD	1769	88	D	S4	F	105	12.1	43	9	4370661680		1200
Coupe												
216	1590	111	P	S4	F	121	9.5	31	14	4270	1680	1090
200 vi	1795	145	P	S4	F	131	7.8	28	15	4270	1680	1200
400												
414	1396	103	P	S4	F	115	11.0	32	9	4490	1700	1120
416	1590	111	P	S4	F	118	10.0	30	12	4490	1700	1125
420	1994	136	P	S4	F	124	9.0	27	13	4490	1700	1265
420 D	1994	86	D	S4	F	105	13.0	47	10	4490	1700	1240
420 Di	1994	105	D	S4	F	115	10.4	50	11	4490	1700	1250
600												
618	1997	115	P	S4	F	121	10.5	27	10	4650	1720	1275
620	1997	131	P	S4	F	125	9.5	25	11	4650	1720	1290
620ti	1994	197	P	S4	F	143	7.0	25	17	4650	1720	1355
623	2259	158	P	S4	F	134	8.2	25	14	4650	1720	1330
620 Diesel	1994	104	D	S4	F	115	10.8	48	10	4650	1720	1300
800												
820 i	1994	136	P	S4	F	125	9.6	25	11	4650	1720	1340
820 Turbo	1994	200	P	S4	F	143	7.3	26	17	4650	1720	1395
825 Diesel	2498	118	D	S4	F	118	10.5	36	11	4650	1720	1445
825	2497	175	P	V6	F	135	8.2	25	14	4650	1720	1390
SAAB												
900												
2.0	1985	140	P	S4	F	124	10.5	23	10	4635	1710	1295
2.3	2290	155	P	S4	F	131	9.5	24	12	4635	1710	1295
2.5	2498	175	P	V6	F	140	8.7	23	14	4635	1710	1350
2.0 Turbo	1985	190	P	S4	F	143	8.0	25	15	4635	1710	1360
9000 CS												
2.0i	1985	130	P	S4	F	124	11.0	23	14	4790	1765	1360
2.0 EcoPower	1985	150	P	S4	F	131	9.5	24	14	4790	1765	1360
2.3 EcoPower	2290	170	P	S4	F	137	8.0	24	15	4790	1765	1360
2.3 Turbo	2290	200	P	S4	F	146	7.2	24	17	4790	1765	1360
2.3 Turbo Aero	2290	225	P	S4	F	149	6.7	23	17	4790	1765	1375
3.0	2962	211	P	V6	F	143	7.6	22	17	4790	1765	1460
SATURN												
SC												
SC1	1901	100	P	S4	F	94	10.5	28	-	4575	1710	1050
SC2	1901	124	P	S4	F	100	9.0	27	-	4575	1710	1080
SEAT												
Marbella												
0.9	903	41	P	S4	F	84	19.2	35	-	3475	1500	720
Ibiza												
1.05	1043	45	P	S4	F	86	22.3	35	2	3815	1640	895
1.4	1391	60	P	S4	F	98	13.9	31	4	3815	1640	895
1.6	1598	74	P	S4	F	106	12.7	29	6	3815	1640	980
1.8 16v	1781	130	P	S4	F	130	9.0	27	12	3815	1640	1030

Technical Data

Model	cc	bhp	F	●	D	mph	A	mpg	UK	⬍	↔	▼
SEAT cont'd												
2.0	1984	114	P	S4	F	121	10.3	29	10	3815	1640	1015
1.9 Diesel	1896	63	D	S4	F	103	16.0	42	6	3815	1640	980
Cordoba												
1.4	1391	60	P	S4	F	98	13.9	31	6	4110	1640	1020
1.6	1598	74	P	S4	F	106	12.5	31	7	4110	1640	1030
2.0	1984	114	P	S4	F	123	10.1	29	11	4110	1640	1130
1.9 Diesel	1896	63	D	S4	F	103	16.5	42	7	4110	1640	1110
1.9 TD	1896	75	D	S4	F	106	15.0	42	8	4110	1640	1015
Toledo												
1.6	1595	75	P	S4	F	106	13.3	32	8	4320	1660	985
1.8	1781	90	P	S4	F	113	11.8	27	9	4320	1660	1050
2.0	1984	115	P	S4	F	122	9.4	26	11	4320	1660	1080
2.0 16V	1984	150	P	S4	F	134	8.4	27	15	4320	1660	1150
1.9 Diesel	1896	68	D	S4	F	103	15.5	41	8	4320	1660	1050
1.9 TD	1896	75	D	S4	F	106	14.9	37	9	4320	1660	1080
1.9 TDi	1896	90	D	S4	F	114	12.7	51	11	4320	1660	1130
Alhambra												
2.0	1984	115	P	S4	F	111	15.4	24	12	4615	1810	1560
1.9 TD	1896	90	D	S4	F	100	19.3	36	12	4615	1810	1670
SKODA												
Felicia												
1.3	1289	54	P	S4	F	90	17.0	34	4	3855	1635	920
1.3	1289	68	P	S4	F	94	14.0	38	4	3855	1635	935
1.6	1598	75	P	S4	F	100	12.0	32	-	3855	1635	965
1.9 D	1896	64	D	S4	F	93	16.0	43	-	3855	1635	995
SPECTRE												
R42	4601	350	P	V8	R	175	4.0	-	-	4178	1854	1250
SSANGYONG												
Musso												
2.3	2295	150	P	S4	4x4	100	-	-	-	4640	1850	1855
3.2	3199	220	P	S6	4x4	120	-	-	-	4640	1850	1890
2.3 Diesel	2299	77	D	S4	4x4	84	-	-	-	4640	1850	1740
2.9 Diesel	2874	94	D	S5	4x4	91	19.7	24	12	4640	1850	1885
SUBARU												
Justy												
1.3	1298	68	P	S4	4x4	96	13.6	33	6	3845	1590	865
Impreza												
1.6	1597	88	P	F4	F	108	12.1	29	10	4350	1690	990
1.6	1597	88	P	F4	4x4	103	13.6	29	10	4350	1690	1070
1.8	1994	116	P	F4	4x4	115	10.0	26	13	4350	1690	1070
2.0 Turbo	1994	208	P	F4	4x4	142	6.6	25	17	4350	1690	1200
Legacy												
2.0i	1994	113	P	F4	4x4	118	10.1	25	14	4595	1695	1060
2.0 Turbo	1994	200	P	F4	4x4	144	6.7	23	-	4595	1695	1375
2.2i	2212	126	P	F4	4x4	121	9.5	24	15	4595	1695	1275
2.5	2457	160	P	F4	4x4	-	-	-	-	4720	1715	1410
SVX												
Coupe	3319	226	P	F6	4x4	146	8.6	19	18	4625	1770	1610
SUZUKI												
Swift												
1.0	993	52	P	S4	F	91	32.4	47	4	3845	1590	730
1.3	1298	67	P	S4	F	101	14.5	43	6	3845	1590	790
1.3 GTi	1298	100	P	S4	F	112	8.6	36	11	3845	1590	770
Baleno												
1.3	1299	86	P	S4	F	106	-	37	1	4195	1690	910
1.6	1590	97	P	S4	F	109	-	33	7	4195	1690	970
1.8	1840	121	P	S4	F	118	10.2	28	-	4195	1690	1000
X-90												
1.6	1590	96	P	S4	4x4	93	-	29	6	3710	1695	1100
Vitara												
1.6 base	1590	74	P	S4	4x4	88	-	31	8	4030	1670	1080
1.6	1590	96	P	S4	4x4	94	-	26	8	4030	1670	1080
2.0 V6	1998	134	P	S4	4x4	99	12.5	22	12	4030	1670	1210
2.0 TD	1998	71	D	S4	4x4	81	-	32	-	4030	1670	1405
TATA												
Gurkha												
2.0 Diesel	1948	68	D	S4	R	80	20.0	31	7	4410	1710	1540
TATRA												
613/5												
3.5 V8	3495	220	P	V8	R	138	7.7	21	-	5130	1800	1810
TOFAS												
Sahin												
1.6	1585	84	P	S4	F	100	13.6	26	-	4310	1640	945
TOYOTA												
Starlet												
1.3	1332	74	P	S4	F	103	11.9	40	4	3740	1635	765
Corolla												
1.3	1332	87	P	S4	F	105	11.8	37	10	4270	1685	945
1.6	1587	113	P	S4	F	116	10.3	33	13	4270	1685	975
1.6 20v	1587	160	P	S4	F	131	-	21	-	4270	1685	1100
2.0 Diesel	1975	71	D	S4	F	103	13.9	42	10	4270	1685	1040
Carina E												
1.6	1587	100	P	S4	F	110	12.9	35	8	4530	1695	1060
1.8	1762	107	P	S4	F	115	11.9	34	10	4530	1695	1060
2.0	1998	133	P	S4	F	122	9.9	29	13	4530	1695	1100
2.0 GTi	1998	175	P	S4	F	134	8.2	26	15	4530	1695	1220
2.0 TD	1975	72	D	S4	F	108	13.0	39	10	4530	1695	1120
Camry												
Data on new model unavailable as we went to press												
Avalon												
3.0	2995	192	P	V6	F	138	9.0	20	-	-	-	-
Paseo												
1.5	1497	90	P	S4	F	114	10.9	36	6	4160	1660	920
MR2												
2.0	1998	154	P	S4	R	137	7.7	27	16	4160	1660	1210
2.0 Turbo	1998	245	P	S4	R	150	-	16	-	4160	1660	1260
Celica												
1.8	1762	114	P	S4	F	124	10.0	32	12	4420	1750	1140
2.0	1998	173	P	S4	F	139	7.9	28	15	4420	1750	1190
2.0 GT-Four	1998	238	P	S4	4x4	153	5.9	24	17	4420	1750	1380
Previa												
2.4	2438	133	P	S4	R	111	11.3	22	13	4765	1800	1625
RAV4												
2.0	1998	129	P	S4	4x4	108	10.7	27	9	4115	1690	1150
Landcruiser Colorado												
3.0 TD	2982	123	D	S4	4x4	99	14.0	21	-	4750	1820	1720
3.4	3378	176	P	V6	4x4	109	11.5	15	-	4750	1820	1810
Landcruiser												
3.0 TD	2982	123	D	S4	4x4	93	16.4	24	12	4885	1690	1660
VX 4.4 auto	4477	202	P	S4	4x4	106	12.4	15	15	4885	1690	2260
VX 4.2 TD	4164	168	D	S4	4x4	106	12.5	26	13	4885	1690	2090
TVR												
Chimaera												
4.0	3950	240	P	V8	R	152	4.8	15	20	4015	1865	1060
4.0 HC	3950	275	P	V8	R	158	4.6	-	20	4015	1865	1060
5.0	4997	290	P	V8	R	162	4.4	11	20	4015	1865	1060
Cerbera												
4.2	4185	360	P	V8	R	165	4.1	-	20	4280	1865	1100
Griffith 500												
5.0	4997	340	P	V8	R	167	4.1	11	20	3890	1945	1060
VAUXHALL												
Corsa												
1.2i	1195	45	P	S4	F	90	18.0	37	3	3730	1610	825
1.2i 'E'	1195	45	P	S4	F	90	18.5	39	3	3730	1610	825
1.4i	1389	60	P	S4	F	96	14.0	33	4	3730	1610	865
1.4i-16v	1389	90	P	S4	F	112	10.5	33	6	3730	1610	905
1.6i-16v	1589	106	P	S4	F	119	9.5	31	11	3730	1610	945
1.5 TD	1488	67	D	S4	F	103	13.0	47	6	3730	1610	935
1.7 Diesel	1686	60	D	S4	F	96	15.5	48	4	3730	1610	910

Technical Data

	cc	bhp	F	●	D	mph	A	mpg	UK	⇕	↔	▼
VAUXHALL cont'd												
Astra												
1.4i HT	1389	60	P	S4	F	99	15.0	32	6	4240	1695	975
1.4i -16v	1389	90	P	S4	F	110	12.0	32	9	4240	1695	975
1.6i	1598	100	P	S4	F	118	10.5	34	10	4240	1695	990
1.8i	1799	115	P	S4	F	129	8.5	29	12	4240	1695	1080
2.0i	1998	136	P	S4	F	137	9.0	28	12	4240	1695	1110
1.7 Diesel	1700	68	D	S4	F	102	14.5	40	8	4240	1695	1030
1.7 TD	1686	82	D	S4	F	107	12.5	41	9	4240	1695	1050
Vectra												
1.6	1598	75	P	S4	F	108	15.5	31	6	4480	1710	1170
1.6 16v	1598	109	P	S4	F	117	12.5	32	7	4480	1710	1200
1.8 16v	1796	124	P	S4	F	126	11.0	31	9	4480	1710	1265
2.0 16v	1998	136	P	S4	F	132	11.0	27	12	4480	1710	1285
2.5 24v	2498	170	P	V6	F	144	8.5	26	15	4480	1710	1370
2.0 Di												
Omega												
2.0i 8v	1998	115	P	S4	R	121	12.0	24	12	4790	1785	1400
2.0i 16v	1998	136	P	S4	R	130	10.0	26	13	4790	1785	1425
2.5 TD	2498	130	D	S6	R	124	11.0	31	13	4790	1785	1525
2.5 V6	2498	170	P	V6	R	139	8.5	24	14	4790	1785	1510
3.0 V6	1962	210	P	V6	R	149	8.3	22	16	4790	1785	1575
Tigra												
1.4i-16v	1389	90	P	S4	F	118	10.5	33	11	3920	1605	980
1.6i-16v	1598	106	P	S4	F	126	9.5	31	13	3920	1605	1000
Calibra												
2.0i 8v	1998	115	P	S4	F	127	10.0	25	13	4490	1690	1215
2.0i 16v	1998	136	P	S4	F	133	8.0	27	14	4490	1690	1250
2.5 V6	2498	167	P	V6	F	147	7.3	25	17	4490	1690	1325
Turbo	1997	206	P	S4	4x4	152	6.4	24	17	4490	1690	1375
Sintra												
2.2i	2198	141	P	S4	F	118	12.8	26	-	4670	1830	1620
3.0i	2962	201	P	V6	F	126	10.9	21	-	4670	1830	1710
Frontera												
2.0i Sport	1998	115	P	S4	4x4	98	14.5	22	10	4190	1780	1695
2.2i Estate	2198	136	P	S4	4x4	100	12.7	24	10	4690	1780	1805
2.5 TD	2500	115	D	S4	4x4	97	14.8	25	10	4690	1780	1830
Monterey												
3.2 V6	3165	177	P	V6	4x4	106	11.5	16	12	4700	1745	1795
3.1 TD	3059	114	D	S4	4x4	94	16.6	24	12	4700	1745	1880
VENTURI												
210	2458	210	P	V6	R	153	6.9	21	-	4240	1840	1255
260	2849	260	P	V6	R	169	5.3	18	-	4240	1840	1255
300	2975	280	P	V6	R	175	-	-	-	4240	1840	1250
400 GT	2975	408	P	V6	R	181	-	-	-	4240	1840	1150
VOLKSWAGEN												
Polo												
1.0	1043	50	P	S4	F	94	18.5	42	3	3715	1655	880
1.4	1391	60	P	S4	F	100	14.9	37	4	4160	1655	880
1.6	1598	75	P	S4	F	107	12.2	35	6	4160	1655	990
1.6 16v	1595	100	P	S4	F	116	10.7	38	6	4160	1655	990
1.9 Diesel	1896	64	P	S4	F	100	16.1	44	6	4160	1655	930
Golf												
1.4	1391	60	P	S4	F	98	16.3	34	6	4020	1695	1000
1.6	1598	75	P	S4	F	104	14.0	32	7	4020	1695	1015
1.8	1781	75	P	S4	F	113	12.1	29	8	4020	1695	1015
2.0	1984	115	P	S4	F	124	10.1	27	14	4020	1695	1110
2.0 16v	1984	150	P	S4	F	134	8.7	26	16	4020	1695	1165
2.8 VR6	2792	174	P	V6	F	141	7.6	22	16	4020	1695	1210
1.9 Diesel	1896	64	D	S4	F	97	17.6	44	7	4020	1695	1085
1.9 Diesel Ecomatic	1896	64	D	S4	F	97	17.6	61	7	4020	1695	1085
1.9 TD	1896	75	D	S4	F	103	15.4	39	8	4020	1695	1095
1.9 TDi	1896	90	D	S4	F	110	12.8	52	8	4020	1695	1120
Vento												
1.6	1598	75	P	S4	F	104	14.6	30	7	4380	1695	1105
1.8	1781	75	P	S4	F	112	12.5	29	9	4380	1695	1105
2.0	1984	115	P	S4	F	123	10.4	26	11	4380	1695	1160
2.8 VR6	2792	174	P	V6	F	140	7.8	22	16	4380	1695	1265
1.9 Diesel	1896	64	D	S4	F	97	18.1	42	8	4380	1695	1150

	cc	bhp	F	●	D	mph	A	mpg	UK	⇕	↔	▼
VOLKSWAGEN cont'd												
1.9 TD	1896	75	D	S4	F	103	15.7	39	9	4380	1695	1160
1.9 TDI	1896	90	D	S4	F	110	12.8	52	9	4380	1695	1205
Passat												
1.6	1595	100	P	S4	F	120	12.3	29	-	4675	1740	1275
1.8	1781	125	P	S4	F	129	10.9	28	-	4675	1740	1350
1.8 Turbo	1781	150	P	S4	F	139	8.7	29	-	4675	1740	1355
2.3	2327	150	P	V5	F	138	9.1	25	-	4675	1740	1410
2.8	2771	193	P	V6	4x4	149	7.6	22	-	4675	1740	1525
1.9 TDi	1896	90	D	S4	F	115	13.9	46	-	4675	1740	1370
1.9 TDi 110	1896	110	D	S4	F	123	11.7	47	-	4675	1740	13705
Sharan												
2.0	1984	115	P	S4	F	111	15.4	24	11	4615	1810	1560
2.8	2792	174	P	V6	F	124	10.0	21	14	4615	1810	1670
1.9 TD	1896	90	D	S4	F	100	15.4	37	11	4615	1810	1670
VOLKSWAGEN Brazil/Mexico												
Gol												
1.0	997	50	P	S4	F	91	18.4	39	-	3810	1640	895
1.6	1595	76	P	S4	F	105	12.8	35	-	3810	1640	900
1.8	1781	96	P	S4	F	112	10.9	26	-	3810	1640	910
2.0	1984	141	P	S4	F	128	8.5	33	-	3810	1640	1010
1.6 Diesel	1588	54	D	S4	F	83	-	43	-	3810	1640	940
Beetle												
1.6	1584	44	P	F4	R	80	18.0	30	-	4060	1550	820
VOLVO												
440/460												
1.6	1596	83	P	S4	F	109	12.8	29	6	4310	1680	980
1.8	1794	89	P	S4	F	112	11.8	26	8	4310	1680	980
2.0	1998	108	P	S4	F	119	10.0	27	10	4310	1680	1030
1.7 Turbo	1721	120	P	S4	F	121	9.1	26	11	4310	1680	1050
1.9 TD	1870	89	D	S4	F	114	11.8	38	8	4310	1680	980
S40/V40												
1.8	1731	115	P	S4	F	121	10.5	27	10	4480	1720	1200
2.0	1948	140	P	S4	F	130	9.3	26	11	4480	1720	1220
850												
2.0 10v	1984	126	P	S5	F	121	11.7	24	11	4710	1760	1370
2.0 20v	1984	143	P	S5	F	125	10.9	22	12	4710	1760	1370
2.3 T5	2319	225	P	S5	F	149	7.4	21	16	4710	1760	1420
2.3 T5-R	2319	240	P	S5	F	155	6.9	22	17	4710	1760	1420
2.5 10v	2435	140	P	S5	F	127	10.3	23	13	4710	1760	1400
2.5 20v	2435	170	P	S5	F	131	9.3	22	13	4710	1760	1400
2.5 TDi	2461	140	D	S5	F	125	10.1	39	12	4710	1760	1460
940												
2.3 LPT	2316	135	P	S4	R	118	10.3	23	14	4870	1750	1395
2.3 Turbo	2316	155	P	S4	R	127	9.3	22	16	4870	1750	1435
2.4 TD	2383	122	D	S4	R	115	11.6	30	15	4870	1750	1445
960												
2.5	2499	170	P	S6	R	130	9.7	20	15	4870	1750	1500
3.0	2922	204	P	S6	R	130	9.1	18	17	4870	1750	1500
WESTFIELD												
1.6	1597	105	P	S4	R	108	6.3	-	-	4165	1625	-
1.8	1796	130	P	S4	R	118	6.5	-	-	4165	1625	-
S8	3950	200	P	V8	R	138	4.3	-	-	4165	1625	-
ZAZ												
Tavria												
1.1	1091	50	P	S4	F	81	18.0	29	-	3710	1550	770
1.3	1288	65	P	S4	F	94	14.5	35	-	3710	1550	770

New Car Prices

The table below gives the UK list prices of new cars as the Guide went to press in early September 1996. Inevitably prices will change over the coming months; for a completely up-to-date listing look in the centre section of the weekly Auto Express magazine.

ALFA ROMEO	
45	
.6ie 3dr	£11,594
.6ie L 3dr	£12,628
.7ie 16v 3dr	£13,660
.0 Cloverleaf 3dr	£14,883
46	
.6 5dr	£12,110
.6 L 5dr	£13,103
.7 16v 5dr	£14,135
.0 ti 5dr	£15,393
55	
.8-16v 4dr	£14,991
.0-16v 4dr	£16,626
.5 V6 4dr	£18,464
64	
.0 Super 4dr	£18,475
.0 Super Lusso 4dr	£20,929
.0 V6 Super 4dr	£24,305
.0 V6 S' Lusso 4dr	£27,467
.0 V6 Cloverleaf 4dr	£30,427
Spider/GTV	
Spider 2.0 Twin Spark	£22,000
Spider 2.0 TS Lusso	£24,500
GTV 2.0 Twin Spark	£19,950
GTV 2.0 TS Lusso	£22,450

ASIA MOTORS	
Rocsta	
.8 DX Soft Top 2dr	£9,500
.8 DX Hard Top 3dr	£9,800
.2 Dsl Soft Top 2dr	£9,500
.2 Dsl DX Soft Top	£9,500
.2 Dsl DX Hard Top	£9,800

ASTON MARTIN	
DB7	
.2 Coupe	£82,500
.2 Volante	£89,950
Virage	
.3 Volante	£147,862
Coupe	
.3 V8	£139,500
Vantage	
.3 Vantage	£177,600

AUDI	
A3	
.6	£13,796
.6 Sport	£15,378
.6 SE	£15,980
.8	£15,388
.8 Sport	£16,972
.8 SE	£17,575
.8 Turbo Sport	£17,860
A4	
.6 4dr	£16,166
.6 SE 4dr	£17,777
.8 4dr	£18,123
.8 SE 4dr	£19,995
.8 T 4dr	£20,951
.8 T Sport 4dr	£22,588
.9 TDi 4dr	£18,181
.9 TDi SE 4dr	£20,053
.9 TDi 110 4dr	£19,366
.9 TDi 110 SE 4dr	£21,238
.9 TDi 110 quattro	£23,569
6 V6 4dr	£21,367
.6 V6 SE 4dr	£23,143
.8 V6 4dr	£24,625
.8 V6 SE 4dr	£26,399
.8 V6 quattro	o
vant 1.6	£17,065
vant 1.6 SE	£18,852
vant 1.8	£19,023
vant 1.8 SE	£21,070
vant 1.8 T	£21,850
vant 1.8 T Sport	£23,662
vant 1.9 TDi	£19,080
vant 1.9 TDi SE	£21,127
vant 1.9 TDi 110	£20,265
vant 1.9 TDi 110 SE	£22,132
vant 1.9 TDi quattro	£24,645
vant 2.6 V6	£22,267
vant 2.6 V6 SE	£24,217

AUDI cont'd	
Avant 2.8 V6	£25,524
Avant 2.8 V6 SE	£27,474
Avant 2.8 V6 quattro	£29,330
80 Cabriolet	
2.0E	£21,648
2.6E	£25,409
2.8E	£30,296
A6	
1.8 4dr	£18,939
1.8 SE 4dr	£20,877
1.9 TDi 4dr	£20,204
1.9 TDi SE 4dr	£22,143
2.5 TDi 4dr	£21,851
2.5 TDi SE 4dr	£23,790
2.5 TDi 140bhp 4dr	£23,500
2.5 TDi SE 140bhp	£25,439
2.5 TDi 140 quattro	£27,516
2.6 4dr	£22,420
2.6 SE 4dr	£24,265
2.8 4dr	£25,590
2.8 SE 4dr	£27,435
2.8 quattro	£29,950
S6 4dr	£37,500
1.8 Est	£20,525
1.8 Est SE	£22,616
1.9 TDi Est	£21,790
1.9 TDi SE Est	£23,882
2.5 TDi Est	£23,438
2.5 TDi SE Est	£25,529
2.5 TDi 140bhp Est	£25,086
2.5 TDi SE 140 Est	£27,178
2.5 TDi quattro Est 5dr	£29,202
2.6 Est	£24,006
2.6 SE Est	£26,004
2.8 Est	£27,178
2.8 SE Est	£29,174
2.8 quattro Est 5dr	£31,689
S6 Est	£39,239
A8	
2.8 auto 4dr	£35,950
2.8 Sport auto 4dr	£39,480
3.7 auto 4dr	£43,350
3.7 Sport auto 4dr	£46,880
4.2 quattro auto 4dr	£50,274
4.2 qu' Sport auto 4dr	£53,803

BENTLEY	
Brooklands 4dr	£106,866
Turbo R LWB 4dr	£143,268
Continental R	£193,429
Azure	£222,526

BMW	
3-Series	
316i **Compact** 3dr	£13,925
318ti 3dr	£15,960
318tds 3dr	£15,450
316i **Saloon** 4dr	£16,160
316i SE 4dr	£17,630
318i 4dr	£17,480
318i SE 4dr	£18,920
318tds 4dr	£17,340
318tds SE 4dr	£18,780
325tds 4dr	£21,880
325tds SE 4dr	£24,340
328i 4dr	£24,200
328i SE 4dr	£26,100
M3 Evolution 4dr	£37,460
318tds **Touring**	£18,200
318i	£18,370
320i	£23,000
325tds	£24,320
328i	£26,880
316i **Coupe**	£17,450
318iS	£19,650
323i	£23,140
328i	£24,770
328i Sport	£27,900
M3 Evolution	£37,460
318i **Convertible**	£23,730
320i	£26,200
328i	£30,740
M3 Evolution	£42,980

BMW cont'd	
5-Series	
520i 4dr	£23,550
520i SE 4dr	£25,150
523i 4dr	£25,490
523i SE 4dr	£26,990
525tds 4dr	£25,350
525tds SE 4dr	£26,950
528i SE 4dr	£29,990
535i 4dr	£35,950
540i 4dr	£45,520
7-Series	
728i 4dr	£36,300
735i 4dr	£43,090
740i 4dr	£49,950
740iL 4dr	£57,100
750i 4dr	£67,710
750iL 4dr	£72,840
8-Series	
840Ci	£56,850

BRISTOL	
Blenheim	£109,980

CATERHAM	
Seven	
1.6	£17,850
1.6 SS 6-speed	£20,200
2.0 HPC	£21,400
2.0 JPE	£35,000
C21	
1.6	£18,750
1.6 SS 6-speed	£21,100

CHRYSLER	
Neon	
2.0 LE 4dr	£11,595
2.0 LX 4dr	£12,995
Jeep Wrangler	
2.5 Sport 3dr	£13,895
4.0 Sport 3dr	£15,495
4.0 Sahara 3dr	£15,995
Jeep Cherokee	
2.5 Sport 5dr	£16,995
4.0 Sport 5dr	£18,995
4.0 Ltd 5dr	£21,495
4.0 Ltd SE 5dr	£24,295
2.5TD Sport 5dr	£18,195
2.5TD Sport LE 5dr	£18,495
2.5TD Limited 5dr	£21,495
2.5TD Limited SE 5dr	£24,295
Jeep Grand Cherokee	
4.0 5dr	£28,995
5.2 V8 (lhd) 5dr	£29,995
New Yorker	
3.5 V6 4dr	£33,600
Viper	
8.0 (lhd)	£62,250

CITROEN	
AX	
1.0 Debut 3dr	£6,350
1.1 Memphis 3dr	£6,910
1.5 D Debut 3dr	£7,200
1.0 Debut 5dr	£6,770
1.1 Memphis 5dr	£7,325
1.5 D Debut 5dr	£7,620
1.5 D Memphis 5dr	£8,265
Saxo	
1.1 LX 3dr	£7,350
1.1 SX 3dr	£7,900
1.4 SX 3dr	£8,520
1.4 VSX 3dr	£9,390
ZX	
1.4 Elation 5dr	£10,070
1.4 Dimension 5dr	£10,335
1.9 Elation D 5dr	£11,055
1.9 Elation TD 5dr	£11,925
1.9 Dimension TD 5dr	£12,190
1.9 Elation D Est	£11,780
1.9 Dimension D Est	£12,045
1.9 Avantage TD Est	£13,315
1.9 Aura TD Est	£14,195
Xantia	
1.6i 8v LX 5dr	£13,110

CITROEN cont'd	
1.8i 16v LX 5dr	£13,165
1.8i 16v SX 5dr	£14,880
2.0i 16v LX 5dr	£13,730
2.0i 16v SX 5dr	£15,445
2.0i 16v VSX 5dr	£17,890
2.0i Turbo Activa 5dr	£18,985
1.9 D 5dr	£12,375
1.9 D LX 5dr	£13,705
1.9 D SX 5dr	£15,420
1.9 TD 5dr	£12,910
1.9 TD LX 5dr	£14,410
1.9 TD SX 5dr	£15,790
1.9 TD VSX 5dr	£18,260
1.8 8v LX Estate	£14,270
2.0 16v SX Est	£16,445
2.0 8v VSX Est	£19,730
2.0 Turbo VSX Est	£19,330
1.9 TD Est	£13,910
1.9 TD LX Est	£15,240
1.9 TD SX Est	£16,995
1.9 TD VSX Est	£19,260
XM	
2.0i 16v SX 5dr	£18,435
2.0i Turbo SX 5dr	£19,680
2.0i Turbo VSX 5dr	£21,480
2.0i Turbo Exclusive	£25,465
2.1 TD SX 5dr	£20,340
2.1 TD VSX 5dr	£22,125
2.1 TD Exclusive 5dr	£27,285
2.5 TD VSX 5dr	£23,350
2.5 TD Exclusive 5dr	£27,200
3.0 V6 Excl 5dr	£30,150
2.0i 16v SX Est	£18,435
2.0i Turbo VSX Est	£22,020
2.1 TD VSX Est	£22,665
2.5 TD VSX	£23,885
3.0 SEi V6 Est auto	£30,690
Synergie	
2.0i LX	£16,200
2.0i SX	£18,130
1.9 TD LX	£17,200
1.9 TD SX	£19,130
1.9 D VSX	£23,090

DAEWOO	
Nexia	
1.5 GLi 3dr	£8,795
1.5 GLXi 3dr	£9,995
1.5 GLi 5dr	£9,295
1.5 GLXi 5dr	£10,495
1.5 GLi 4dr	£9,295
1.5 GLXi 4dr	£10,495
Espero	
1.5 GLXi 4dr	£11,495
1.8 CDi 4dr	£11,995
2.0 CDXi 4dr	£12,995

DAIHATSU	
Charade	
1.3 Xi 3dr	£6,995
1.3 LXi 5dr	£8,295
1.3 GLXi 5dr	£9,295
1.3 GLXi SE 5dr	£9,995
1.5 GLXi 4dr	£9,695
1.5 GLXi SE 4dr	£10,395
Hijet	
1.0	£7,995
1.0 SE	£8,995
Fourtrak Independent	
2.8 TDS Est 3dr	£13,995
2.8 TDL Est 3dr	£14,995
2.8 TDL-SE Est 3dr	£16,995
2.8 TDX Est 3dr	£17,995
Sportrak	
1.6 Xi Est 3dr	£9,995
1.6 Xi-SE Est 3dr	£11,495
1.6 ELXi Est 3dr	£12,995

FERRARI	
F355	
3.5	£92,657
3.5 GTS	£94,671
3.5 Spider	£98,700
456	£161,143

FIAT	
Cinquecento	
900 3dr	£5,587
900 SX 3dr	£6,038
1.1 Sporting 3dr	£6,487
Punto	
1.1 55 S 3dr	£6,928
1.1 55 SX 3dr	£7,536
1.1 55 SX 6 spd 3dr	£8,313
1.2 75 SX 3dr	£8,253
1.2 75 ELX 3dr	£9,189
1.4 GT 3dr	£12,025
1.6 Sporting 3dr	£9,928
1.7 D S 3dr	£7,687
1.7 TD SX 3dr	£8,884
1.1 55 S 5dr	£7,304
1.1 55 SX 5dr	£7,911
1.2 60 SX Selecta 5dr	£9,192
1.2 75 SX 5dr	£8,628
1.2 75 ELX 5dr	£9,563
1.6 90 ELX 5dr	£9,928
1.7 D S 5dr	£8,062
1.7 TD SX 5dr	£9,259
1.6 ELX Cabrio	£13,728
Bravo	
1.4 S 3dr	£9,761
1.4 SX 3dr	£10,738
1.6 SX 3dr	£11,284
1.8 HLX 3dr	£12,780
Brava	
1.4 S 5dr	£10,145
1.4 SX 5dr	£11,007
1.6 SX 5dr	£11,550
1.6 ELX 5dr	£12,780
1.8 ELX 5dr	£13,532
barchetta	
1.8	£14,260
Coupé	
2.0 16v	£17,923
2.0 16v Turbo	£19,860
Ulysse	
2.0 S	£15,895
2.0 EL	£18,995
1.9 TD S	£16,645
1.9 TD EL	£19,745

FORD	
Fiesta	
1.3 Encore 3dr	£7,645
1.3 LX 3dr	£8,645
1.25-16v LX 3dr	£9,165
1.25-16v Si 3dr	£10,095
1.4-16v Si 3dr	£10,510
1.8 D Encore 3dr	£8,350
1.8 D LX 3dr	£9,350
1.3 Encore 5dr	£8,085
1.3 LX 5dr	£9,085
1.3 Ghia 5dr	£10,380
1.25-16v LX 5dr	£9,605
1.25-16v Si 5dr	£10,535
1.25-16v Ghia 5dr	£10,900
1.25-16v Ghia X 5dr	£12,585
1.4-16v Si 5dr	£10,950
1.4-16v Ghia 5dr	£11,315
1.4-16v Ghia X 5dr	£12,060
1.8 D Encore 5dr	£8,790
1.8 D LX 5dr	£9,790
1.8 D Ghia 5dr	£10,680
Escort	
1.3 Encore 3dr	£9,995
1.4 Encore 3dr	£10,450
1.4 L 3dr	£10,960
1.4 LX 3dr	£11,345
1.6 L 16v auto 3dr	£12,180
1.6 Si 16v 3dr	£12,330
1.8 Si 16v 3dr	£12,930
1.8 TDsl Encore 3dr	£10,950
1.3 Encore 5dr	£10,445
1.4 Encore 5dr	£10,900
1.4 L 5dr	£11,410
1.4 LX 5dr	£11,795
1.6 L 16v 5dr	£11,675
1.6 LX 16v 5dr	£12,505
1.6 Si 16v 5dr	£12,780
1.6 Ghia 16v 5dr	£13,235

FORD cont'd	
1.6 Ghia X 16v 5dr	£15,715
1.8 LX 16v 5dr	£12,505
1.8 Si 16v 5dr	£13,380
1.8 Ghia 16v 5dr	£13,235
1.8 Ghia X 16v 5dr	£14,760
1.8 TD Encore 5dr	£11,400
1.8 TD 70 PS L 5dr	£11,910
1.8 TD 70 PS LX 5dr	£12,295
1.8 TD 90 PS LX 5dr	£13,545
1.8 TD 90 PS Ghia 5dr	£14,275
1.4 LX 4dr	£11,795
1.6 L 16v 4dr	£11,675
1.6 LX 16v 4dr	£12,505
1.6 Ghia 16v 4dr	£13,235
1.8 LX 16v 4dr	£12,505
1.8 Ghia 16v 4dr	£13,235
1.8 TD 70 PS L 4dr	£11,910
1.8 TD 70 PS LX 4dr	£12,295
1.8 TD 90 PS LX 4dr	£13,545
1.8 TD 90 PS Ghia 4dr	£14,275
1.4 Encore Est	£11,700
1.6 L 16v Est	£12,475
1.6 LX 16v Est	£13,305
1.6 Si 16v Est	£13,580
1.6 Ghia 16v Est	£14,305
1.8 LX 16v Est	£13,305
1.8 Si 16v Est	£14,180
1.8 Ghia 16v Est	£14,035
1.8 TDsl 70 PS Encore	£12,200
1.8 TDsl 70 PS L Est	£12,710
1.8 TDsl 70 PS LX Est	£13,095
1.8 TD 90 PS LX Est	£14,345
1.8 TD 90 PS Ghia Est	£15,075
1.6 Calypso	£14,495
1.8 Cabrio Ghia	£15,995
Mondeo	
1.6i Aspen 5dr	£12,810
1.6i LX 5dr	£13,895
1.8i LX 5dr	£13,895
1.8i GLX 5dr	£15,295
1.8 TD Aspen 5dr	£13,430
1.8 TD LX 5dr	£14,515
1.8 TD GLX 5dr	£15,915
1.8 TD Ghia 5dr	£16,300
1.8 TD Ghia X 5dr	£18,900
2.0i LX 5dr	£13,895
2.0i GLX 5dr	£15,295
2.0i Si 5dr	£15,295
2.0i Ghia 5dr	£15,995
2.0i Ghia X 5dr	£18,900
2.5 V6 Si 5dr	£17,870
2.5 V6 Ghia 5dr	£18,570
2.5 V6 Ghia X 5dr	£21,475
1.6i LX 4dr	£13,895
1.8i LX 4dr	£13,895
1.8i GLX 4dr	£15,295
1.8 TD LX 4dr	£14,515
1.8 TD GLX 4dr	£15,915
1.8 TD Ghia 4dr	£16,300
1.8 TD Ghia X 4dr	£18,900
2.0i LX 4dr	£13,895
2.0i GLX 4dr	£15,295
2.0i Si 4dr	£15,295
2.0i Ghia 4dr	£15,995
2.0i Ghia X 4dr	£18,900
2.5 V6 Si 4dr	£17,870
2.5 V6 Ghia 4dr	£18,570
2.5 V6 Ghia X 4dr	£21,475
1.8i LX Est	£15,025
1.8i GLX Est	£16,425
1.8 TD Aspen Est	£14,560
1.8 TD LX Est	£15,645
1.8 TD GLX Est	£17,045
1.8 TD Ghia Est	£17,430
1.8 TD Ghia X Est	£20,030
2.0i LX Est	£15,025
2.0i GLX Est	£16,425
2.0i Ghia Est	£17,125
2.0i Ghia X Est	£20,030
2.5 V6 Ghia Est	£19,700
2.5 V6 X Ghia Est	£22,605
Scorpio	
2.0 8v Ghia 4dr	£18,700
2.3 16v Ghia 4dr	£19,800

FORD cont'd

2.3 16v Ghia X 4dr	£21,400
2.3 16v Ultima 4dr	£23,500
2.9 24v Ghia 4dr	£23,500
2.9 24v Ghia X 4dr	£25,100
2.9 24v Ultima 4dr	£27,200
2.5 TD Ghia 4dr	£20,800
2.5 TD Ghia X 4dr	£22,400
2.5 TD Ultima 4dr	£23,500
2.0 8v Ghia estate	£18,700
2.3 16v Ghia estate	£19,800
2.3 16v Ghia X estate	£21,400
2.3 16v Ultima estate	£23,500
2.9 24v Ghia estate	£23,500
2.9 24v Ghia X estate	£25,100
2.9 24v Ultima estate	£27,200
2.5 TD Ghia estate	£20,800
2.5 TD Ghia X estate	£22,400
2.5 TD Ultima estate	£23,500
Probe	
2.0 16v	£17,845
2.5 V6 24v	£21,210
Galaxy	
2.0 Aspen 5st	£16,720
2.0 Aspen 7st	£17,220
2.0 GLX 7st	£18,590
2.0 Ghia 6st	£20,320
2.8 GLX 7st	£21,390
2.8 Ghia 6st	£22,370
1.9 TD Aspen 5st	£17,720
1.9 TD Aspen 7st	£18,220
1.9 TD GLX 7st	£19,590
1.9 TD Ghia 6st	£21,320
Maverick	
2.4 GLS 3dr	£17,180
2.4 GLS 5dr	£19,930
2.7 TD GLS 3dr	£18,230
2.7 TD GLS 5dr	£20,980

FSO

Caro	
1.5 GLXi 5dr	£5,999
1.9 GLD 5dr	£6,999

HONDA

Civic	
1.4i 3dr	£11,975
1.5 LSi 3dr	£13,585
1.6 VTi 3dr	£16,085
1.4i (entry) 5dr	£12,175
1.4i 5dr	£13,085
1.5i VTEC-E 5dr	£13,540
1.6i LS 5dr	£14,585
1.6i SR 5dr	£15,735
1.5 LSi 4dr	£14,385
1.6 VTi 4dr	£16,885
1.6i LS Coupe	£13,705
1.6 SR VTEC Coupe	£15,235
CRX	
1.6 ESi	£18,085
Accord	
1.8i 4dr	£14,785
2.0i S 4dr	£16,085
2.0i LS 4dr	£17,285
2.0i ES 4dr	£19,975
2.2i VTEC 4dr	£21,940
2.0 TDi 4dr	£18,785
2.0i LS Coupe	£18,140
2.2i ES Coupe	£20,975
2.0i LS Aerodeck	£17,585
2.2i ES Aerodeck	£20,685
Shuttle	£23,585
Prelude	
2.0	£18,885
2.3	£22,695
2.2 VTEC	£23,885
New Legend	
3.5 Saloon 4dr	£33,585
NSX	
3.0	£69,475
3.0-T	£72,725

HYUNDAI

Accent	
1.3 Coupe 3dr	£6,999
1.3 Coupe Si 3dr	£8,199
1.3 LSi 5dr	£8,199
1.5 GSi 5dr	£9,199
1.5 GLSi 5dr	£9,999
1.3 LSi 4dr	£8,199
1.5 GLSi 4dr	£9,999
Lantra	
1.6 LSi 4dr	£9,999
1.6 GLSi 4dr	£11,499
1.8 Si 4dr	£12,299
1.8 CD 4dr	£13,749
1.6 LSi Estate	£10,999
1.6 GLSi Estate	£11,999
1.8 CD Estate	£14,249
Coupe	
2.0	£14,999
2.0 SE	£16,499
Sonata	
2.0 GLX 4dr	£13,399
2.0 CD 4dr	£15,499
3.0 V6 auto 4dr	£17,999

ISUZU

Trooper	
3.2 V6 SWB 3dr	£17,999
3.2 V6 SWB Duty 3dr	£18,999
3.2 V6 SWB Cit 3dr	£21,749
3.1 TD SWB 3dr	£17,999
3.1 TD SWB Duty 3dr	£18,999
3.1 TD SWB Cit 3dr	£21,749
3.2 V6 LWB 5dr	£20,399
3.2 V6 LWB Duty 5dr	£21,499
3.2 V6 LWB Citation	£24,799
3.1 TD LWB 5dr	£20,399
3.1 TD LWB Duty 5dr	£21,499
3.1 TD LWB Cit 5dr	£24,799

JAGUAR

XJ Saloon	
XJ6 3.2 4dr	£30,100
XJ6 3.2 lwb 4dr	£32,850
XJ Executive 4dr	£34,970
Sovereign 3.2 4dr	£39,650
Sovereign 3.2 lwb 4dr	£42,400
Sovereign 4.0 4dr	£44,400
Sovereign 4.0 lwb 4dr	£47,150
XJ6 3.2 Sport 4dr	£31,300
XJ6 4.0 Sport 4dr	£36,650
XJR 4.0 4dr	£48,000
XJ12 6.0 4dr	£54,450
XJ12 6.0 lwb 4dr	£57,200
Daimler 6 lwb 4dr	£54,000
Daimler 6 Century	£63,400
Daimler Double 6	£65,500
Daimler Century 6.0	£71,000

KIA

Pride	
1.3i Start 3dr	£5,489
1.3i LX 3dr	£6,249
1.3i Start 5dr	£5,989
1.3i LX 5dr	£6,749
Mentor	
1.6 Start 4dr	£8,489
1.6 SLX 4dr	£8,949
1.6 GLX 4dr	£9,849
1.6 GLX SE 4dr	£10,249
1.6 GLX Executive 4dr	£11,399
Sportage	
2.0 SLX 5dr	£13,299
2.0 GLX 5dr	£14,199
2.0 GLX SE 5dr	£15,199
2.0 GLX Executive 5dr	£16,199
2.0 GLX SE Exec 5dr	£17,199

LADA

Samara	
1.1 3dr	£4,495
1.3 S 3dr	£5,435
1.5 S 3dr	£5,845
1.3 S 5dr	£5,845
1.5 S 5dr	£6,345
1.3 S 4dr	£5,845
1.5 S 4dr	£6,345
Riva	
1.5 E 4dr	£4,495
1.5 E Est	£5,495
Niva	
1.7i Hussar 3dr	£7,995
1.7i Cossack 3dr	£9,995

LAMBORGHINI

Diablo	
5.7 SV	£124,950
5.7 Coupe	£148,000
5.7 VT	£162,000
5.7 Roadster	£175,000

LANDROVER

Defender	
2.5 County TDi 3dr	£18,876
2.5 County TDi 5dr	£21,550
Discovery	
2.0 Mpi 3dr	£19,150
2.5 Tdi 3dr	£20,030
3.9 V8i 3dr	£20,050
2.0 Mpi 5dr	£21,275
2.0 Mpi S 5dr	£22,510
2.5 Tdi 5dr	£22,125
2.5 Tdi S 5dr	£23,390
2.5 Tdi XS 5dr	£26,475
2.5 Tdi ES 5dr	£29,605
3.9 V8i S 5dr	£23,860
3.9 V8i XS 5dr	£26,475
3.9 V8i ES 5dr	£30,075
Range Rover	
2.5 DT 5dr	£34,515
2.5 DT SE 5dr	£34,515
4.0 V8 5dr	£39,100
4.0 V8 SE 5dr	£39,100
4.6 HSE 5dr	£47,150

LEXUS

GS300	
3.0 4dr	£33,395
3.0 Sport 4dr	£36,390
LS400 4dr	£45,995

LISTER

Storm	£219,725

LOTUS

Elise	
1.8	£19,950
Esprit	
S4	£48,995
S4S	£52,995
V8	£58,750

MARCOS

Mantara	
Mantara Coupe 4.0	£27,929
Mantara Coupe 5.0	£33,924
LM Coupe 4.0	£31,695
LM Coupe 5.0	£37,690
Mantara Spyder 4.0	£28,553
Mantara Spyder 5.0	£34,548
LM Spyder 4.0	£32,318
LM Spyder 5.0	£38,313

MASERATI

Ghibli	£44,931
Ghibli GT	£47,549
Quattroporte 4dr	£57,669

MAZDA

121	
1.3 GXi 3dr	£7,885
1.25-16v ZXi 3dr	£9,485
1.8 D DXi 3dr	£8,340
1.3 GXi 5dr	£8,340
1.25-16v ZXi 5dr	£9,940
323	
1.5i LXi 5dr	£11,995
1.5i GXi 5dr	£12,595
1.8i Exec 5dr	£13,995
2.0i V6 SXi 5dr	£16,995
626	
1.8i LXi 4dr	£12,595
1.8i LXi 5dr	£12,695
1.8i GXi 5dr	£13,995
2.0i GXi 5dr	£14,750
2.0i Executive 5dr	£17,995
Xedos 6	
1.6i 4dr	£15,995
2.0i V6 Sport 4dr	£18,995
2.0i V6 SE 4dr	£22,195
Xedos 9	
2.5 V6 auto 4dr	£26,995
MX-3	
1.6i auto	£14,795
1.8i V6	£16,995
MX-5	
1.6i	£13,495
1.8i	£14,495
1.8i S	£17,595
MX-6	
2.5i	£20,245

McLAREN

F1	£634,500

MERCEDES-BENZ

C-Class	
C180 Classic/Esprit	£19,250
C200 Classic/Esprit	£21,400
C220 Classic/Esprit	£23,300
C230 Classic/Esprit	£23,600
C230K Classic/Esprit	£25,600
C280 Classic/Esprit	£27,250
C220 Classic/Esprit D	£20,800
C250 Classic/Esprit D	£22,850
C250 Classic/Esprit TD	£23,300
C180 Cl'c/Esprit Est	£20,680
C200 Cl'c/Esprit Est	£22,830
C230 Cl'c/Esprit Est	£25,030
C220 Cl'c/Esprit D Est	£22,230
C250 Cl'c/Esprit TD Est	£24,730
Elegance + £2,700; Sport + £3,100	
C36 4dr	£42,400
E-Class	
E200 Classic 4dr	£24,000
E230 Classic 4dr	£26,500
E280 Classic 4dr	£31,900
E320 Classic 4dr	£35,200
E250D Classic 4dr	£25,300
E300D Classic 4dr	£27,450
E200 Classic Est	£26,300
E230 Classic Est	£28,800
Elegance +£2,550; Avant Garde +£2,800	
Old E-Class	
E220 Cabrio	£40,300
E320 Cabrio	£51,650
S-Class	
S280 4dr	£41,000
S320 4dr	£49,200
S420 4dr	£59,900
S500 4dr	£69,300
S320L 4dr	£51,200
S420L 4dr	£62,200
S500L 4dr	£73,400
S600L 4dr	£101,900

MERCEDES-BENZ cont'd

CL-Class	
S420	£69,900
S500	£80,800
S600	£105,400
SL	
SL280	£57,700
SL320	£62,900
SL500	£80,700
SL600	£102,400

MGF

1.8i	£16,396
1.8i VVC	£18,796

MITSUBISHI

Colt	
1.3 GLX 3dr	£9,799
1.6 GLX 3dr	£11,299
1.6 Mirage 3dr	£12,499
Carisma	
1.6 GL 5dr	£10,999
1.6 GLX 5dr	£11,999
1.8 GLX 5dr	£12,299
1.8 GLS 5dr	£13,499
1.6 GL 4dr	£10,999
1.6 GLX 4dr	£11,999
1.8 GLX 4dr	£12,299
1.8 GLS 4dr	£13,499
Galant (1996 model)	
2.0 GLSi DOP 4dr	£17,269
2.0 V6-24v 4dr	£19,469
Space Runner	
1.8	£14,329
1.8 DOP	£15,169
Space Wagon	
2.0 GLXi	£15,999
2.0 GLXi DOP	£16,999
Shogun	
2.5 TD GLX 3dr	£18,879
2.5 TD GLS 3dr	£21,539
2.5 TD GLS DOP 3dr	£23,349
3.0-24v V6 3dr	£22,589
3.0-24v V6 DOP 3dr	£24,399
3.5-24v V6 SE 3dr	£29,759
2.8 TD GLX 5dr	£22,959
2.8 TD GLS 5dr	£26,139
2.8 TD GLS DOP 5dr	£29,239
3.0-24v V6 5dr	£27,059
3.0-24v V6 DOP 5dr	£30,159
3.5-24v V6 SE 5dr	£36,079
3000 GT	£43,999

MORGAN

4/4	
1.8 2-seater	£17,831
Plus 4	
2.0 2-seater	£21,896
2.0 4-seater	£23,529
Plus 8	
3.9 2-seater	£28,470

NISSAN

Micra	
1.0 Shape 3dr	£7,195
1.0 S 3dr	£8,550
1.0 GX 3dr	£8,550
1.3 GX 3dr	£9,010
1.3 SR 3dr	£10,350
1.0 Shape 5dr	£7,630
1.0 GX 5dr	£8,985
1.3 GX 5dr	£9,445
1.3 SLX 5dr	£10,595
Almera	
1.4 Equation 3dr	£9,455
1.4 GX 3dr	£10,995
1.4 Si 3dr	£10,995
1.6 SRi 3dr	£12,495
2.0 GTi 3dr	£13,995
1.4 Equation 5dr	£9,965
1.4 GX 5dr	£11,495
1.4 Si 5dr	£11,495
1.6 GX 5dr	£12,125
1.6 SRi 5dr	£12,995
1.6 SLX 5dr	£12,995
2.0 D GX 5dr	£12,125
1.6 GX 4dr	£12,125
1.6 SLX 4dr	£12,995
2.0 D GX 4dr	£12,125
Primera (1996 model)	
1.6 Precision 5dr	£11,708
1.6 LX 5dr	£13,340
1.6 SRi 5dr	£13,790
2.0 SLX 5dr	£15,530
2.0 SRi 5dr	£14,585
2.0i SE 5dr	£17,340
2.0E GT 5dr	£17,340
2.0D LX 5dr	£13,340
1.6 LX 4dr	£13,340
1.6 SRi 4dr	£13,790
2.0 SLX 4dr	£15,30
2.0 SRi 4dr	£14,585
2.0i SE 4dr	£17,340
2.0E GT 4dr	£17,340
2.0D LX 4dr	£13,340
1.6 LX Est	£14,375
2.0 LX Est	£15,130

NISSAN cont'd

2.0 SLX Est	£16,270
2.0D LX Est	£14,275
QX	
2.0 V6 S 4dr	£16,995
2.0 V6 SE 4dr	£19,295
3.0 V6 SE auto 4dr	£23,745
3.0 V6 SEL auto 4dr	£27,1455
SX	
2.0 Turbo	£19,650
2.0 Turbo Touring	£22,250
Serena	
1.6 GX	£14,457
2.0 Excursion	£16,545
2.3 GXD	£14,457
2.3 SLXD	£15,897
Terrano II	
2.4 SR 3dr	£16,995
2.4 SR Sport 3dr	£18,695
2.4 SE 5dr	£19,495
2.7 TDi S 3dr	£15,995
2.7 TDi SR 3dr	£17,995
2.7 TDi SE 3dr	£17,995
2.7 TDi SE 5dr	£20,495
2.7 TDi SR 5dr	£20,495
2.7 TDi SE Touring 5dr	£22,495
Patrol	
GR 2.8 SLX TD 3dr	£20,495
GR 2.8 SLX TD 5dr	£25,945
GR 4.2 SE 5dr	£27,145

PEUGEOT

106	
1.1 XN 3dr	£7,145
1.1 XL 3dr	£7,745
1.4 XR 3dr	£9,395
1.5 XND 3dr	£7,645
1.5 XLD 3dr	£8,245
1.5 XRD 3dr	£9,395
1.1 XN 5dr	£7,545
1.1 XL 5dr	£8,145
1.4 XR 5dr	£9,795
1.5 XND 5dr	£8,045
1.5 XLD 5dr	£8,645
1.5 XRD 5dr	£9,795
306	
1.4 XN 3dr	£10,095
1.4 XL 3dr	£11,050
1.6 XS 3dr	£12,130
2.0 XSi 3dr	£14,110
2.0 GTI-6 3dr	£16,675
1.9 XND 3dr	£10,745
1.9 D Turbo 3dr	£12,970
1.4 XN 5dr	£10,500
1.4 XL 5dr	£11,455
1.4 XR 5dr	£12,105
1.6 XR 5dr	£12,305
1.8 XT 5dr	£13,600
2.0 XSi 5dr	£14,515
1.9 XND 5dr	£11,150
1.9 XLD 5dr	£11,855
1.9 XRD 5dr	£12,505
1.9 XLDT 5dr	£12,455
1.9 XRDT 5dr	£13,105
1.9 XTDT 5dr	£14,000
1.9 D Turbo 5dr	£13,375
1.4 SN Sedan 4dr	£10,500
1.6 SL Sedan 4dr	£11,880
1.8 SR Sedan 4dr	£13,025
2.0 ST Sedan 4dr	£14,380
1.9 SND Sedan 4dr	£11,150
1.9 D Sedan 4dr	£10,760
1.9 SLDT Sedan 4dr	£12,530
1.9 SRDT Sedan 4dr	£13,475
1.9 STDT Sedan 4dr	£14,380
2.0 Cabrio	£18,895
2.0 Roadster	£19,995
406	
1.8 L 4dr	£12,785
1.8 LX 4dr	£13,695
2.0 LX 4dr	£13,995
2.0 SRi 4dr	£15,355
2.0 GLX 4dr	£15,595
2.0 Executive 4dr	£18,945
2.0 Turbo SRi 4dr	£16,345
2.0 Turbo Executive 4dr	£19,170
1.9 TD L 4dr	£13,435
1.9 TD LX 4dr	£14,345
1.9 TD GLX 4dr	£15,995
1.9 TD Executive 4dr	£18,935
2.1 TDLX 4dr	£15,195
2.1 TDGLX 4dr	£16,795
2.1 TDExecutive 4dr	£19,735
605	
2.0 SRTi 4dr	£21,550
2.1 SRTD 4dr	£22,270
3.0 SVE 4dr	£28,625
806	
2.0 SL	£16,450
2.0 SR	£17,990
2.0 SV	£22,250
1.9 TD SLD	£17,450
1.9 TD SRD	£18,990
1.9 TD SVD	£23,250

PORSCHE

911	
Carrera Coupe	£61,250
Carrera Targa	£65,950
Turbo	£99,950
Carrera Cabrio	£67,445
Carrera 4 2dr	£64,450
Carrera 4S 2dr	£74,795
Carrera 4 Cabrio 2dr	£70,695

PROTON

Compact	
1.3 LSi 3dr	£8,999
1.5 GLSi 3dr	£9,999
1.6 SEi 3dr	£11,999
Persona	
1.5 GLSi 5dr	£10,650
1.6 XLi 5dr	£12,350
1.8 SEi 5dr	£13,650
1.8 SDi 5dr	£12,350
1.5 GLSi 4dr	£10,699
1.6 XLi 4dr	£11,999
1.8 SEi 4dr	£13,299
1.8 SDi 4dr	£11,999

RENAULT

Clio	
1.2 RL 3dr	£7,490
1.2 RN 3dr	£9,005
1.4 RT 3dr	£10,245
1.8 RSi 3dr	£12,160
1.9 RLD 3dr	£8,075
1.2 RL 5dr	£7,940
1.2 RN 5dr	£9,455
1.4 RT 5dr	£10,695
1.8 Baccara 5dr	£14,750
1.9 RLD 5dr	£8,525
1.9 RND 5dr	£9,805
1.9 RTD 5dr	£10,695
Megane	
1.4e RN 5dr	£10,690
1.4e RT 5dr	£11,370
1.6e RT 5dr	£11,855
1.6e RXE 5dr	£12,905
1.9 D RN 5dr	£11,110
1.9 D RT 5dr	£11,840
1.9 TD RT 5dr	£12,870
1.9 TD RXE 5dr	£13,545
2.0 RXE 5dr	£14,320
1.6 Coupe	£11,855
2.0 Coupe	£14,320
2.0 16v Coupe	£16,450
Laguna	
1.8 RN 5dr	£11,660
1.8 RT 5dr	£12,720
2.0 RT 5dr	£13,405
2.0 RXE 5dr	£15,315
2.0 Executive 5dr	£17,410
2.0 RTi 16v 5dr	£15,665
2.2D RN 5dr	£12,595
2.2D RT 5dr	£13,720
2.2D RXE 5dr	£15,565
2.2TD RT 5dr	£15,960
2.2TD RXE 5dr	£18,000
2.2TD Executive 5dr	£19,900
3.0 V6 auto 5dr	£20,260
3.0 V6 Baccara 5dr	£21,470
1.8 RN Estate	£12,760
2.0 RT Estate	£14,505
2.0 RXE 16v	£17,370
2.2D RN Estate	£13,745
2.2D RT Estate	£14,870
2.2TD RT Estate	£17,110
2.2TD RXE Estate	£19,425
Safrane (1996 model)	
2.0i 5dr	£17,250
2.2 Vi Executive 5dr	£19,850
2.5 TD Exec 5dr	£21,250
3.0 V6i 5dr	£28,175
Espace (1996 model)	
2.0 Helios 5 seat	£16,755
2.0 Helios	£17,315
2.0 Alize	£18,990
2.2 Executive	£22,385
2.1 Helios TD 5 seat	£17,755
2.1 Helios TD	£18,315
2.1 Alize TD	£20,390
2.1 Executive TD	£21,985
2.9 Executive	£26,375

ROLLS-ROYCE

Silver Dawn 4dr	£118,558
Silver Spur 4dr	£135,243

ROVER

Mini (1996 model)	
1.3i Sprite 2dr	£6,846
1.3i Mayfair 2dr	£8,086
1.3i Cooper 2dr	£8,386
1.3i Cabrio	£12,586
100	
111 Knightsbridge 3dr	£7,186
111 Knightsbridge SE 3dr	£7,78[illegible]
114 SLi 3dr	£8,986
114 GTa 3dr	£9,286
115 K'bridge SED 3dr	£8,286
111 Knightsbridge 5dr	£7,586

ROVER cont'd

111 K'bridge SE 5dr £8,186
114 SLi 5dr £9,396
114 GSi 5dr £9,986
114 Cabrio 5dr £12,586
115 K'bridge SED 5dr £8,686

200

214i 3dr £10,986
214 Si 3dr £12,086
216 Si 3dr £13,086
200 vi 3dr c £15,000
220 D 3dr £11,386
214i 5dr £11,486
214 Si 5dr £12,586
216 Si 5dr £13,586
216 SLi 5dr £14,786
220 D 5dr £11,886
220 SD 5dr £13,086
220 SDi 5dr £14,186

400

414i 5dr £12,986
414 Si 5dr £13,786
416i 5dr £13,786
416 Si 5dr £14,686
416 SLi 5dr £15,786
420i 5dr £14,686
420 Si 5dr £14,586
420 SLi 5dr £16,386
420 GSi 5dr £17,086
420 D 5dr £14,286
420 SD 5dr £15,186
420 SDi 5dr £15,686
420 SLDi 5dr £16,486
420 GSDi 5dr £17,486
416i 4dr £14,286
416 Si 4dr £15,186
416 SLi 4dr £15,986
420i 4dr £14,786
420 Si 4dr £15,686
420 SLi 4dr £16,486
420 GSi 4dr £17,386
420 D 4dr £14,786
420 SD 4dr £15,686
420 SDi 4dr £16,186
420 SLDi 4dr £16,986
420 GSDi 4dr £17,786

Coupe/Cabrio/Tourer

1.6 Coupe £17,086
1.8 VVC Coupe £19,286
1.6 Cabrio £16,086
1.6 SE Cabrio £17,586
1.6 Tourer £16,086
1.8 TD Tourer £16,186
1.8 VVCTourer £17,986

600

618i 4dr £15,441
618 Si 4dr £16,441
620 Si 4dr £17,586
620 SLi 4dr £19,086
620 GSi 4dr £20,586
620ti 4dr £20,586
620 SDi 4dr £17,586
620 SLDi 4dr £19,086
620 GSDi 4dr £20,586
623 GSi 4dr £24,586

800

820 i 5dr £18,586
820 Si 5dr £20,086
820 SLi 5dr £22,086
820 Vitesse 5dr £22,586
825 Di 5dr £21,086
825 SDi 5dr £22,586
825 SLDi 5dr £24,586
825 Si 5dr £22,586
825 SLi 5dr £24,586
Sterling 5dr £29,086
820 i 4dr £18,586
820 Si 4dr £20,086
820 SLi 4dr £22,086
820 Vitesse 4dr £22,586
825 Di 4dr £21,086
825 SDi 4dr £22,586
825 SLDi 4dr £24,586
825 Si 4dr £22,586
825 SLi 4dr £24,586
Sterling 4dr £29,086
820 Coupe Turbo £26,586
800 Coupe auto £30,586

SAAB

900

2.0 3dr £14,402
2.0 S 3dr £16,252
2.3 S Coupe 3dr £18,302
2.0 S Turbo Coupe 3dr £20,202
2.5 V6 SE Coupe 3dr £22,602
2.0 SE Turbo C'pe 3dr £22,802
2.0 5dr £14,902
2.0 S 5dr £16,752
2.3 S 5dr £18,802
2.0 XS 5dr £18,402
2.5 V6 SE 5dr £23,102
2.0 SE Turbo 5dr £23,302
2.0 S Conv' £21902
2.3 SE Conv' £24,902
2.5 SE V6 Conv' £28,202
2.0 SE Turbo Conv' £28,402

SAAB cont'd

9000 CS

CS 2.0i 5dr £18,402
CS 2.0 EcoPower 5dr £19,602
CS 2.3 EcoPower 5dr £20,802
CS 2.3 Turbo Eco 5dr £23,402
CS 3.0 V6 5dr £26,702
CSE 2.0 EcoPower 5dr £23,002
CSE 2.3 EcoPower 5dr £24,302
CSE 2.3 Turbo Eco 5dr £26,002
CSE 3.0 V6 5dr £28,002
Aero 2.3 Turbo 5dr £30,152

9000 CD

CD 2.0i 4dr £18,402
CD 2.0 EcoPower 4dr £20,152
CD 2.3 EcoPower 4dr £20,802
CD 2.3 Turbo Eco 4dr £23,402
CD 3.0 V6 4dr £26,702
CDE 2.0 EcoPower 4dr £23,002
CDE 2.3 EcoPower 4dr £24,302
CDE 2.3 Turbo Eco 4dr £26,002
CD 3.0 V6 Griffin 4dr £30,152

SEAT

Ibiza

1.05 CL 3dr £7,485
1.4 CLS 3dr £8,195
1.4 Salsa 3dr £9,195
1.8 GTi 16v 3dr £13,150
2.0 GTi 3dr £12,250
1.9D CLS 3dr £8,750
1.4 CLS 5dr £8,595
1.4 Salsa 5dr £9,595
1.6 S 5dr £11,275
1.9D CLS 5dr £9,150
1.9D Salsa 5dr £10,195

Cordoba

1.4 CLS 4dr £9,785
1.6 CLX 4dr £10,775
1.6 GLX 4dr £12,065
2.0 GT 4dr £13,950
1.9D CLS 4dr £10,095
1.9D CLX 4dr £10,930
1.9 TD GLX 4dr £13,110
1.6 SX Coupe £12,595

Toledo

1.6 SE 5dr £12,095
1.8 SE 5dr £13,095
2.0 Sport 5dr £15,095
2.0 GT 16V 5dr £17,295
1.9 D SE 5dr £12,295
1.9 TD SE 5dr £12,595
1.9 TDi SE 5dr £14,095
1.9 TDi SXE 5dr £15,095

Alhambra

1.9 TDi SE £17,210
2.0 SE £16,455

SKODA

Felicia

1.3 LX 5dr £6,175
1.3 LXi 5dr £6,585
1.3 LXi Plus 5dr £7,050
1.3 GLXi 5dr £7,923
1.6 GLXi 5dr £8,499
1.3 LXi Estate £7,585
1.3 LXi Plus Estate £8,050
1.3 GLXi Estate £8,923
1.6 GLXi Estate £9,499

SSANGYONG

Musso

2.9D Standard 5dr £15,999
2.9D SE 5dr £17,499
2.9D GSE 5dr £20,999
3.2 GX220 5dr c£25,000

SUBARU

Justy

4wd 1.3 GX 3dr £8,599
4wd 1.3 GX 5dr £8,999

Impreza

2wd 1.6 GL 5dr £11,249
4wd 1.6 GL 5dr £12,499
4wd 2.0 GL 5dr £13,249
4wd 2.0 Sport 5dr £14,249
4wd 2.0 Turbo 5dr £18,499
2wd 1.6 LX 4dr £9,999
2wd 1.6 GL 4dr £10,749
4wd 2.0 GL 4dr £12,749
4wd 2.0 Sport 4dr £13,749
4wd 2.0 Turbo 4dr £17,999

Legacy

4wd 2.0i GL 4dr £13,648
4wd 2.0i GLS 4dr £15,249
4wd 2.2i GX 4dr £16,999
4wd 2.0i GL Est £14,999
4wd 2.0i GLS Est £16,748
4wd 2.2i GX Est £18,999

SVX

Coupe £30,499

SUZUKI

Swift

1.0 GLS 3dr £6,295
1.3 GLS 3dr £7,145
1.3 GTi 3dr £9,750

SUZUKI cont'd

1.0 GC 5dr £6,295
1.3 GX 5dr £7,295

Baleno

1.6 GS 3dr £8,995
1.6 GLX 4dr £9,995
1.6 GLX Executive 4dr £11,595
1.8 £11,595

X-90

1.6 2wd £9,950
1.6 4wd £10,050

Vitara

1.6 JLX Soft Top 2dr £11,895
1.6 JLX SE Soft Top 2dr £12,695
1.6 JLX Est 3dr £11,895
1.6 JLX SE Est 3dr £12,695
1.6 JX Est 5dr £12,695
2.0 V6 5dr £15,775
2.0 TD 5dr £15,775

TATA

Gurkha

2.0D £9,995
2.0D SE £11,995

TOYOTA

Starlet

1.3 Sportif 3dr £7,599
1.3 CD 3dr £9,549
1.3 Sportif 5dr £8,519
1.3 CD 5dr £9,999

Corolla

1.3 Sportif 3dr £9,999
1.3 GS 3dr £11,439
1.3 CD 3dr £12,639
1.6 Si 3dr £13,299
1.3 Sportif 5dr £10,509
1.3 GS 5dr £11,949
1.6 CD 5dr £12,919
1.6 CDX 5dr £14,279
2.0 GSD 5dr £12,429
1.6 CD 4dr £13,249
2.0 GSD Est £13,079

Carina E

1.6 S 5dr £11,849
1.6 GS 5dr £13,139
1.8 GS 5dr £13,299
1.8 CD 5dr £14,275
1.8 GLi 5dr £14,409
1.8 CDX 5dr £16,425
2.0 GLi 5dr £15,319
2.0 CDX 5dr £17,325
2.0 GS TD 5dr £13,899
2.0 GL TD 5dr £15,089
1.8 GS 4dr £13,149
1.8 GLi 4dr £14,259
1.8 CDX 4dr £16,275
2.0 GLi auto 4dr £16,029
1.8 GS Est £13,975
2.0 GLi Est £15,989
2.0 GS TD Est £14,575

Camry (1996 model)

2.2i 16v 4dr £19,185
3.0i V6 GX 4dr £24,919
2.2i 16v est £20,370
3.0i V6 est £24,529

Paseo

1.5 ST £12,485
1.5 Si £13,940

MR2

2.0 GT £20,995
2.0 GT T–bar £22,615

Celica

1.8 ST £17,699
2.0 GT £22,521
2.0 Cabrio £27,975
2.0 GT-Four 3dr £31,289

Previa

2.4 GS £18,766
2.4i GL £22,579
2.4i GX £26,839

RAV4

2.0 3dr £13,326
2.0 GX 3dr £14,670
2.0 GX 5dr £16,245

Landcruiser Colorado

3.0 TD GS 3dr £19,990
3.0 TD GX 3dr £23,990
3.0 TD GX 5dr £26,990
3.0 TD VX 5dr £31,990
3.4 V6 VX 5dr £32,990

Landcruiser

4.2 TD GS 5dr £30,166
4.2 TD VX 5dr £39,869
4.4 VX 5dr £39,549

TVR

Chimaera

4.0 £29,450
4.0 HC £30,450
5.0 £34,595

Cerbera

4.2 £37,000

Griffith 500

5.0 £34,595

VAUXHALL

Corsa

1.2i Merit 3dr £7,375
1.2i LS 3dr £8,175
1.4i LS 3dr £8,650
1.4i-16v Sport 3dr £10,600
1.7D Merit 3dr £8,000
1.7D LS 3dr £9,125
1.2i Merit 5dr £7,775
1.2i LS 5dr £8,575
1.2i GLS 5dr £9,725
1.4i LS 5dr £9,050
1.4i-16v GLS 5dr £10,775
1.4i-16v CDX 5dr £11,80
1.7D Merit 5dr £8,400
1.7D LS 5dr £9,525
1.5TD GLS 5dr £11,050

Astra

1.4i HT Merit 3dr £10,450
1.4i HT LS 3dr £11,345
1.6i-16v Sport 3dr £12,745
1.7TD Merit 3dr £10,950
2.0i-16v Sport 3dr £13,995
1.4i HT Merit 5dr £10,900
1.4i HT LS 5dr £11,795
1.4i-16v LS 5dr £12,395
1.4i-16v GLS 5dr £13,145
1.6i Merit 5dr £11,300
1.6i LS 5dr £12,395
1.6i-16v GLS 5dr £13,495
1.6i-16v Sport 5dr £13,195
1.6i-16v CDX 5dr £14,995
1.7TD Merit 5dr £11,400
1.7TD LS 5dr £12,495
1.7TDS LS 5dr £13,295
1.7TDS GLS 5dr £13,745
2.0i-16v Sport 5dr £14,445
2.0i-16v CDX 5dr £16,245
1.6i LS 4dr £12,395
1.6i-16v GLS 4dr £13,495
1.7TD LS 4dr £12,495
1.7TDS GLS 4dr £13,745
1.4i HT Merit Est £11,650
1.4i-16v LS Est £13,145
1.6i Merit Est £12,050
1.6i LS Est £13,145
1.6i-16v GLS Est £14,245
1.6i-16v CDX Est £15,745
1.7TD Merit Est £12,145
1.7TD LS Est £13,245
1.7TDS LS Est £14,045
1.7TDS GLS Est £14,495
2.0i-16v Sport Est £15,195
2.0i-16v CDX Est £16,995
1.6i Colour Conv £15,220
1.8i Colour Conv £17,900

Vectra

1.6 Envoy 5dr £12,875
1.6-16v LS 5dr £13,825
1.6-16v GLS 5dr £15,650
1.8-16v LS 5dr £14,145
1.8-16v GLS 5dr £15,650
2.0-16v LS 5dr £14,290
2.0-16v GLS 5dr £16,150
2.0-16v SRi 5dr £16,150
2.0-16v CDX 5dr £19,550
2.5-24v V6 GLS 5dr £17,150
2.5-24v V6 SRi 5dr £17,695
2.5-24v V6 CDX 5dr £21,045
1.7 TD Envoy 5dr £13,505
1.7 TD LS 5dr £14,495
1.7 TD GLS 5dr £15,900
1.7 TD CDX 5dr £18,950
1.6-16v LS 4dr £13,825
1.6-16v GLS 4dr £15,650
1.8-16v LS 4dr £14,145
1.8-16v LS 4dr £15,650
2.0-16v LS 4dr £14,290
2.0-16v GLS 4dr £16,150
2.0-16v SRi 4dr £16,150
2.0-16v CDX 4dr £19,550
2.5-24v V6 GLS 4dr £17,150
2.5-24v V6 SRi 4dr £17,695
2.5-24v V6 CDX 4dr £21,045
1.7 TD Envoy 4dr £13,505
1.7 TD LS 4dr £14,495
1.7 TD GLS 4dr £15,900
1.7 TD CDX 4dr £18,950

Omega

2.0i Select 4dr £18,250
2.0i 16v GLS 4dr £19,400
2.0i 16v CD 4dr £21,550
2.5 TD Select 4dr £20,500
2.5 TD GLS 4dr £21,400
2.5 TD CD 4dr £23,550
2.5 TD CDX 4dr £26,600
2.5 TD Elite 4dr £29,500
2.5 V6 Select 4dr £19,450
2.5 V6 GLS 4dr £20,400
2.5 V6 CD 4dr £22,550
2.5 V6 CDX 4dr £25,600
3.0 V6 Elite 4dr £29,500
2.0i Select Est £19,000
2.0i 16v GLS Est £20,150
2.0i 16v CD Est £22,300
2.5 TD Select Est £21,250
2.5 TD GLS Est £22,150

VAUXHALL cont'd

2.5 TD CD Est £24,300
2.5 TD CDX Est £27,350
2.5 TD Elite Est £29,500
2.5 V6 Select Est £20,200
2.5 V6 GLS Est £21,150
2.5 V6 CD Est £23,300
2.5 V6 CDX Est £26,350
3.0 V6 Elite Est £29,500

Tigra

1.4i-16v £11,995
1.6i-16v £13,006

Calibra

2.0i 16v £18,700
2.5 V6 £20,700

Frontera

2.0i Sport 3dr £14,995
2.0i Sport S 3dr £15,885
2.5 TDS Sport 3dr £16,950
2.8 TDS Sport S 3dr £17,845
2.2i Est 5dr £18,795
2.8TD Est 5dr £19,995

Monterey

3.1 TD RS 3dr £23,300
3.1 TD LTD 5dr £25,295
3.1 TD Diamond 5dr £27,185
3.2 V6 LTD 5dr £26,570
3.2 V6 Diamond 5dr £28,460

VOLKSWAGEN

Polo

1.0 L 3dr £7,335
1.4 L 3dr £8,235
1.4 CL 3dr £9,285
1.6 CL 3dr £9,785
1.9D L 3dr £8,985
1.9D CL 3dr £9,585
1.0 L 5dr £7,795
1.4 L 5dr £8,695
1.4 CL 5dr £9,745
1.6 CL 5dr £10,245
1.6 GL 5dr £10,985
1.9D L 5dr £9,445
1.9D CL 5dr £10,045
1.6 L 4dr £9,635
1.6 CL 4dr £10,245
1.6 CL100 4dr £10,735
1.9 L SDI 4dr £10,025
1.9 CL SDI 4dr £10,635

Golf

1.4 L 3dr £10,250
1.9 L D 3dr £10,650
2.0 GTi 3dr £13,880
2.0 GTi 16v 3dr £15,780
2.8 VR6 3dr £18,715
2.8 VR6 H'line 3dr £21,215
1.4 L 5dr £10,720
1.6 CL 5dr £11,680
1.8 GL 5dr £13,385
1.9 L D 5dr £11,120
1.9 CL TD 5dr £12,180
1.9 CL TDi 5dr £13,135
1.9 GL TDi 5dr £14,350
2.0 GTi 5dr £14,350
2.0 GTi 16v 5dr £16,250
2.8 VR6 5dr £19,185
2.8 VR6 H'line 5dr £21,685
1.8 L Est £11,890
1.8 CL Est £12,950
1.9 L D Est £11,890
1.9 CL TD Est £12,950
1.9 CL TDi Est £13,905
1.9 GL TDi Est £15,120
2.0 GL Est £15,120
1.8 Cabrio 75bhp £14,235
1.8 Cabrio 90bhp £16,335
2.0 A'garde Cabrio £18,335

Vento

1.8 CL 4dr £12,180
1.9 CL TD 4dr £12,180
1.9 CL TDI 4dr £13,135
1.9 GL TDI 4dr £14,350
2.0 GL 4dr £14,350
2.8 VR6 4dr £19,185

Passat (1996 model)

1.8 L 4dr £11,995
1.8 CL 4dr £13,745
1.9 L TD 4dr £11,995
1.9 CL TDi 4dr £14,495
1.9 GL TDi 4dr £15,945
2.0 CL 4dr £14,495
2.0 GL 4dr £15,595
1.8 L Est £12,995
1.8 CL Est £14,745
1.9 L TD Est £12,995
1.9 CL TDi Est £15,495
1.9 GL TDi Est £16,945
2.0 CL Est £15,495
2.0 GL Est £16,945

Sharan

1.9 CL TDi £16,983
1.9 GL TDi £19,688
1.9 Carat TDi £21,338
2.0 CL £16,224
2.0 GL £18,929
2.0 Carat £20,579
2.8 VR6 GL £21,779

VOLKSWAGEN cont'd

2.8 VR6 Carat £22,686

VOLVO

440

1.6 5dr £11,570
1.8 5dr £11,870
1.9 TD 5dr £12,370
2.0 5dr £12,370
LE +£425; S, Si +£1280; GS +£1425; SE +£1630; GLT +£3,400; CD +£3910

460

1.6 4dr £11,570
1.8 4dr £11,870
1.9 TD 4dr £12,370
2.0 4dr £12,370
LE +£425; S, Si +£1280; GS +£1425; SE +£1630; GLT +£3,400; CD +£3910

S40/V40

S40 1.8 4dr £14,200
S40 1.8 SE 4dr £15,500
S40 1.8 CD 4dr £17,950
S40 2.0 4dr £14,700
S40 2.0 SE 4dr £16,000
S40 2.0 CD 4dr £18,450
V40 1.8 £14,700
V40 1.8 SE £16,000
V40 1.8 CD £18,450
V40 2.0 £15,200
V40 2.0 SE £16,500
V40 2.0 CD £18,950

850

2.0 10v 4dr £17,950
2.0 20v 4dr £18,950
2.5 10v 4dr £18,650
2.5 20v 4dr £20,050
2.5 TDi 4dr £21,550
2.3 T5 4dr £21,450
2.0 10v Est £18,950
2.0 20v Est £19,950
2.5 10v Est £19,650
2.5 20v Est £21,050
2.5 TDi Est £22,550
2.3 T5 Est £23,450
S + £1,500; SE + £1800; GLT + £3400; CD + 4800
2.3 R 4dr £33,550
2.3 R Est £33,550
2.5 AWD Est 5dr £27,400

940

2.3 LPT 4dr £16,500
2.4 TD 4dr £18,100
2.3 LPT Est £16,700
2.3 Turbo Est £17,600
2.4 TD Est £18,300
S Est + £800; SE Est + £2500; GLE Est + £3800

960

2.5 4dr £20,500
3.0 4dr £22,600
2.5 Est £21,500
3.0 Est £23,600
SE + £1,400; GLE + £2,800; CD + £4,400; Luxury + £3,000-£4,900

WESTFIELD

1600 £11,750
1800 £14,250
130 £16,450
S8 Club £22,950
S8 Euro £25,950

Show notes

Use this page to record useful information from the Motor Show

Model : ______ Stand N° ______	Model : ______ Stand N° ______	Model : ______ Stand N° ______
Model : ______ Stand N° ______	Model : ______ Stand N° ______	Model : ______ Stand N° ______

Show notes

Use this page to record useful information from the Motor Show

Model : ______ Stand N° ______	Model : ______ Stand N° ______	Model : ______ Stand N° ______
Model : ______ Stand N° ______	Model : ______ Stand N° ______	Model : ______ Stand N° ______

Show notes

Use this page to record useful information from the Motor Show

Model : ______ Stand N° ______	Model : ______ Stand N° ______	Model : ______ Stand N° ______
Model : ______ Stand N° ______	Model : ______ Stand N° ______	Model : ______ Stand N° ______